REF
GE
105
.M66
2004
v. 5
AVALON

DISCARD

Form 178 rev. 11-00

Avalon Branch Library
8828 So. Stony Island Ave.
Chicago, ILL 60617

Teen Guides to

Environmental Science

Teen Guides to

Environmental Science

Creating a Sustainable Society
Volume V

John Mongillo
with assistance from Peter Mongillo

Greenwood Press
Westport, Connecticut • London

Library of Congress Cataloging-in-Publication Data

Mongillo, John F.
Teen guides to environmental science / John Mongillo with assistance from Peter Mongillo.
p. cm.
Includes bibliographical references and index.
Contents: v. 1. Earth systems and ecology—v. 2. Resources and energy—v. 3. People and their environments—v. 4. Human impact on the environment—v. 5. Creating a sustainable society.
ISBN 0–313–32183–3 (set : alk. paper)—ISBN 0–313–32184–1 (v. 1 : alk. paper)—
ISBN 0–313–32185–X (v. 2 : alk. paper)—ISBN 0–313–32186–8 (v. 3 : alk. paper)—
ISBN 0–313–32187–6 (v. 4 : alk. paper)—ISBN 0–313–32188–4 (v. 5 : alk. paper)
1. Environmental sciences. 2. Human ecology. 3. Nature–Effect of human beings on. I. Mongillo, Peter A. II. Title.
GE105.M66 2004
333.72—dc22 2004044869

British Library Cataloguing in Publication Data is available.

Copyright © 2004 by John Mongillo

All rights reserved. No portion of this book may be reproduced, by any process or technique, without the express written consent of the publisher.

Library of Congress Catalog Card Number: 2004044869
ISBN: 0–313–32183–3 (set)
0–313–32184–1 (vol. I)
0–313–32185–X (vol. II)
0–313–32186–8 (vol. III)
0–313–32187–6 (vol. IV)
0–313–32188–4 (vol. V)

First published in 2004

Greenwood Press, 88 Post Road West, Westport, CT 06881
An imprint of Greenwood Publishing Group, Inc.
www.greenwood.com

Printed in the United States of America

The paper used in this book complies with the Permanent Paper Standard issued by the National Information Standards Organization (Z39.48–1984).

10 9 8 7 6 5 4 3 2 1

R0402861977

Contents

Acknowledgments

The authors wish to acknowledge and express the contribution of the many nongovernment organizations, corporations, colleges, and government agencies that provided assistance to the authors in the research for this book. The authors are grateful to the Greenwood Publishing Group for permission to excerpt text and photos from *Encyclopedia of Environmental Science*, John Mongillo and Linda Zierdt-Warshaw, and *Environmental Activists*, John Mongillo and Bibi Booth. Both books are excellent references for researching environmental topics and gathering information about environmental activists. Many thanks to those who provided special assistance in reviewing particular topics and offering comments and suggestions: Sara Jones, middle school director for La Jolla Country Day School in San Diego, California; Emily White, teacher of geography and world cultures at the 5th grade level at La Jolla Country Day School, San Diego, California; Lucinda Kramer and John Guido, middle school social studies coordinators, North Haven, Connecticut; Daniel Lanier, environmental professional, and Susan Santone, executive director of Creative Change, Ypsilanti, Michigan.

A special thank you goes to the following people and organizations that provided technical expertise and/or resources for photos and data: Neil Dahlstrom, John Deere & Company; Francine Murphy-Brillon, Slater Mill Historic Site; Lake Worth Public Library, Florida; Pacific Gas & Electric; Energetch; Environmental Justice Resource Center; NASA Johnson Space Center; Seattle Audubon Society; John Onuska, INMETCO; Cathrine Sneed, Garden Project; Denis Hayes, president, Bullitt Foundation; Ocean Robbins, Youth for Environmental Sanity; Maria Perez and Nevada Dove, Friends of McKinley; Juana Beatriz Gutiérrez, cofounder and president of Madres del Este de Los Angeles—Santa Isabel; Mikhail Davis, director, Brower Fund, Earth Island Institute; Randall Hayes, president, Rainforest Action Network; Tom Repine, West Virginia Geologic Survey; Peter Wright and Nancy Trautmann, Cornell University; Mary N. Harrison, University of Florida; and Huanmin Lu, University of Texas, El Paso.

Other sources include Centers for Disease Control and Prevention, Department of Environmental Management, Rhode Island; ChryslerDaimler; Pattonville High School; National Oceanic and

Atmospheric Administration; Chuck Meyers, Office of Surface Mining; U.S. Department of Agriculture; U.S. Fish and Wildlife Service; U.S. Department of Energy; U.S. Environmental Protection Agency; U.S. National Park Service; National Renewable Energy Laboratory; Tower Tech, Inc.; Earthday 2000; Marilyn Nemzer, Geothermal Education Office; U.S. Agricultural Research Service; U.S. Geological Survey; Glacier National Park; Monsanto; CREST Organization; Shirley Briggs, Vortec Corporation; National Interagency Fire Center/Bureau of Land Management; Susan Snyder, Marine Spill Response Corporation; Lisa Bousquet, Roger Williams Park Zoo, Rhode Island; Netzin Gerald Steklis, International National Response Corporation; U.S. Department of the Interior/Bureau of Reclamation; Bluestone Energy Services; OSG Ship Management, Inc.; and Sweetwater Technology.

In addition, the authors wish to thank Hollis Burkhart and Janet Heffernan for their copyediting and proofreading support; Muriel Cawthorn, Hollis Burkhart, and Liz Kincaid for their assistance in photo research; and illustrators Christine Murphy, Susan Stone, and Kurt Van Dexter.

The responsibility of the accuracy of the terms is solely that of the authors. If errors are noticed, please address them to the authors so that corrections can be made in future revisions.

INTRODUCTION

Teen Guides to Environmental Science is a reference tool which introduces environmental science topics to middle and high school students. The five-volume series presents environmental, social, and economic topics to assist the reader in developing an understanding of how human activity has changed and continues to change the face of the world around us.

Events affecting the environment are reported daily in magazines, newspapers, periodicals, newsletters, radio, and television, and on Websites. Each day there are environmental reports about collapsing fish stocks, massive wastes of natural resources and energy, soil erosion, deteriorating rangelands, loss of forests, and air and water pollution. At times, the degradation of the environment has led to issues of poverty, malnutrition, disease, and social and economic inequalities throughout the world. Human demands on the natural environment are placing more and more pressure on Earth's ecosystems and its natural resources.

The challenge in this century will be to reverse the exploitation of Earth's resources and to improve social and economic systems. Meeting these goals will require the participation and commitment of businesses, government agencies, nongovernment organizations, and individuals. The major task will be to begin a long-term environmental strategy that will ensure a more sustainable society.

CREATING A SUSTAINABLE SOCIETY

Sustainable development is a strategy that meets the needs of the present without compromising the ability of future generations to meet their own needs. Many experts believe that for too long, social, economic, and environmental issues were addressed separately without regard to each other. In creating a sustainable society, there needs to be an integration of goals related to economic growth, environmental protection, and social equity. Some of these integrated sustainable goals include the following:

- Improve the quality of human life
- Conserve Earth's diversity

- Minimize the depletion of nonrenewable resources
- Keep within Earth's carrying capacity
- Enable communities to care for their own environments
- Integrate the environment, economy, and human health into decision making
- Promote caretakers of Earth.

OVERVIEW

Teen Guides to Environmental Science provides an excellent opportunity for students to study and focus on the integration of ecological, economical, and social goals in creating a sustainable society. Within the five-volume series, students can research topics from a long list of contemporary environmental issues ranging from alternative fuels and acid rain to wetlands and zoos. Strategies and solutions to solve environmental issues are presented, too. Such topics include soil conservation programs, alternative energy sources, international laws to preserve wildlife, recycling and source reduction in the production of goods, and legislation to reduce air and water pollution, just to name a few.

Major Highlights

- Assists students in developing an understanding of their global environment and how the human population and its technologies have affected Earth and its ecology.
- Provides an interdisciplinary perspective that includes ecology, geography, biology, human culture, geology, physics, chemistry, history, and economics.
- "Raises a student's awareness of a strategy called sustainable development that meets the needs of the present without compromising the ability of future generations to meet their own needs" (Bruntland Commission). The strategy includes a level of economic development that can be sustained in the future while protecting and conserving natural resources with minimum damage to the environment. People concerned about sustainable development suggest that meeting the needs of the future depends on how well we balance social, economic, and environmental objectives—or needs—when making decisions today.
- Presents current environment, social, and economic issues and solutions for preserving wildlife species, rebuilding fish stocks, designing strategies to control sprawl and traffic congestion, and developing hydrogen fuel cells as a future energy source.

- Challenges everyone to become more active in their home, community, and school in addressing environmental problems and discussing strategies to solve them.

ORGANIZATION

Teen Guides to Environmental Science is divided into five volumes.

Earth Systems and Ecology

Volume I begins the discussion of Earth as a system and focuses on ecology—the foundation of environmental science. The major chapters examine ecosystems, populations, communities, and biomes.

Resources and Energy

Currently, fossil fuels drive the economy in much of the world. In Volume II conventional fuels such as petroleum, coal, and natural gas are reported. Other chapters elaborate on nuclear energy, hydrogen energy, wind energy, geothermal energy, solar energy, and natural resources such as soil and minerals, forests, water resources, and wildlife preserves.

People and Their Environments

The history of civilizations, human ecology, and how early and modern societies have interacted with the environment is presented in Volume III. The major chapters highlight the Agricultural Revolution, the Industrial Revolution, global populations, and economic and social systems.

Human Impact on the Environment

Volume IV discusses the causes and the harmful effects of air and water pollution and sustainable solution strategies to control the problems. Other chapters examine the human impact on natural resources and wildlife and discuss efforts to preserve them.

Creating a Sustainable Society

Volume V focuses on the importance of living in a sustainable society in which generations after generations do not deplete the natural resources or produce excessive pollutants. The chapters present an overview of sustainability in producing products, preserving wildlife habitats, developing sustainable communities and transportation systems, and encouraging sustainable management practices in agriculture and commercial fishing. The last chapter in this volume considers the importance of individual activism in identifying and solving environmental problems in one's community.

PROGRAM RESEARCH

The five-volume series represents research from a variety of recurring and up-to-date sources, including newspapers, middle school and high school textbooks, trade books, television reports, professional journals, national and international government organizations, nonprofit organizations, private companies, businesses, and individual contacts.

CONTENT STANDARDS

The series provides a close alignment with the fundamental principles developed and reported in the President's Council on Sustainable Development and the learning outcomes for middle school education standards found in the North American Association for Environmental Education, the National Geography Standards, and the National Science Education Standards.

MAJOR ENVIRONMENTAL TOPICS

The *Teen Guides to Environmental Science* provide terms, topics, and subjects covered in most middle school and high schools environmental science courses. These major topics of environmental science include, but are not limited to:

- Agriculture, crop production, and pest control
- Atmosphere and air pollution
- Ecological economies
- Ecology and ecosystems
- Endangered and threatened wildlife species
- Energy and mineral resources
- Environmental laws, regulations, and ethics
- Oceans and wetlands
- Nonhazardous and hazardous wastes
- Water resources and pollution.

SPECIAL FEATURES

Tables, Figures, and Maps

Hundreds of photos, tables, maps, and figures are ideal visual learning strategies used to enhance the text and provide additional information to the reader.

Vocabulary

The vocabulary list at the end of each chapter provides a definition for a term used within the chapter with which a reader might be unfamiliar.

Marginal Topics

Each chapter contains marginal features which supplement and enrich the main topic covered in the chapter.

Activities

More than 100 suggested student research activities appear at the ends of the chapters in the books.

In-Text References

Many of the chapters have specially marked callouts within the text which refer the reader to other books in the series for additional information. For example, fossil fuels are discussed in Volume V; however, an in-text reference refers the reader to Volume II for more information about the topic.

Websites

A listing of Websites of government and nongovernment organizations is available at the end of each chapter allowing students to research topics on the Internet.

Bibliography

Book titles and articles relating to the subject area of each chapter are presented at the end of each chapter for additional research opportunities.

Appendixes

Four appendixes are included at the end of each volume:

- Environmental Timeline, 1620–2004. To understand the history of the environmental movement, each book provides a comprehensive timeline that presents a general overview of activists, important laws and regulations, special events, and other environmental highlights over a period of more than 400 years.
- Endangered List of U.S. Wildlife Species by State.
- Website addresses by classification..
- Government and nongovernment environmental organizations.

CHAPTER 1

Creating a Sustainable Society

Human activities are placing more and more pressure on Earth's ecosystems and its natural resources. Each day there are environmental reports about collapsing fish stocks, massive wastes of natural resources and energy, soil erosion, deteriorating rangelands, loss of forests, and the increase of air and water pollution. The degradation of the environment has led to issues of poverty, malnutrition, disease, and social and *economic* inequalities throughout the world.

The challenge in this century will be to reverse the exploitation of Earth's resources and improve social and economic systems. Meeting the challenge will require the participation and commitment of businesses, government agencies, nongovernment organizations, and individuals. The major goal will be to create a long-term environmental strategy that will ensure a more *sustainable* society.

DID YOU KNOW?

The word sustainable has Latin roots meaning "to hold up" or "to support from below."

SUSTAINABLE DEVELOPMENT

Most dictionaries define sustainable as "able to sustain," "to keep in existence," "to maintain," or "to endure." A sustainable system would be one that can be continued indefinitely without depleting any of the resources and materials on which it depends. For example, an agricultural system is sustainable if it incorporates the preservation of natural resources, such as soil and water, during the production of food and other supplies for human welfare while generating a profit. Sustainable agriculture also requires the use of technology that does not degrade the potential of the land to produce.

In 1987 the World Commission on Environment and Development called for sustainable development to "meet the needs of the present generation without compromising or forfeiting the ability of future generations to meet their own needs." Five years later, the United Nations Conference on Environment and Development (UNCED) proposed the *Agenda 21* action plan. The plan calls for all governments to adopt a national strategy for sustainable development which includes balancing the needs of our society, economy, and environment, while improving the quality of life for ourselves and for future generations. Since then, sustainable development programs, sometimes known as green plans, have been adopted by many nations throughout the world.

The term sustainable development has met with some opposition. Critics dislike the use of the word development because it is associated with growth, and growth cannot be sustained in a world whose natural resources are finite. They would prefer alternative language such as sustainable communities. Nevertheless, the United Nations and the majority of world governments have accepted the use of the term sustainable development in their policies.

The Goals of Sustainable Development

Many experts believe that, for too long, social, economic, and environmental issues were addressed separately without regard to each other. In creating a sustainable society, there needs to be an integration of goals related to economic growth, environmental protection, and social equity. Some of these integrated sustainable goals include the following:

- Improve the quality of human life
- Conserve Earth's diversity
- Minimize the depletion of nonrenewable resources
- Keep within Earth's *carrying capacity*
- Enable communities to care for their own environments
- Integrate the environment, economy, and human health into decision making
- Promote caretakers of Earth

Social
Equity
Participation
Empowerment
Social Mobility
Cultural Preservation

Economic
Services
Household Needs
Industrial Growth
Agricultural Growth
Efficient Use of Labor

Environment
Biodiversity
Natural Resources
Carrying Capacity
Ecosystem Integrity
Clean Air and Water

FIGURE 1-1 • Social, Economic, and Environmental Goals of a Sustainable Society In creating a sustainable society, there needs to be an integration of goals related to economic growth, environmental protection, and social equity. *Source:* The World Bank Group

CREATING A SUSTAINABLE ENVIRONMENT

The Issues

Many environmentalists confirm that by the year 2100 between 20 and 50 percent of all the wildlife species that lived on Earth since 1900 will become extinct. One environmental study reports that as many as 100 species disappear from Earth each day. Environmentalists are concerned that, as the human population continues growing and altering environments to meet its needs, more organisms will become extinct, thus decreasing Earth's *biodiversity*.

DID YOU KNOW?

A 1992 national opinion survey discovered that only one person in five had ever heard of the loss of biodiversity.

Besides the loss of biodiversity, environmental issues include air pollution, water pollution, and land degradation. Air pollution is a major health problem in the United States and throughout the world. An estimated three million people die each year from the effects of air pollutants. More than one billion people lack access to clean drinking water, while others are forced to use contaminated water. Excessive logging, land degradation by mining, overgrazing of livestock, and poor agricultural management practices are causing the depletion of nutrients in the soil, the wearing away of land to desertlike conditions, and habitat loss.

Creating a More Sustainable Environment, Protecting Biodiversity

A major goal of sustainable development is the conservation of natural resources. This goal involves not only using these resources in such a manner as to prevent their depletion, but also using them in a manner that will not degrade their quality for future generations. For example, freshwater is a natural resource which exists in limited amounts on Earth's surface, but it is needed by most of Earth's organisms for their survival. In many places on Earth, bodies of freshwater have become polluted through the activities of humans or have become unsafe for use because of contamination with pathogens. Water that is potable must be used wisely to ensure that it remains in adequate supply to meet the demands of a growing population. Wastewater also requires special management.

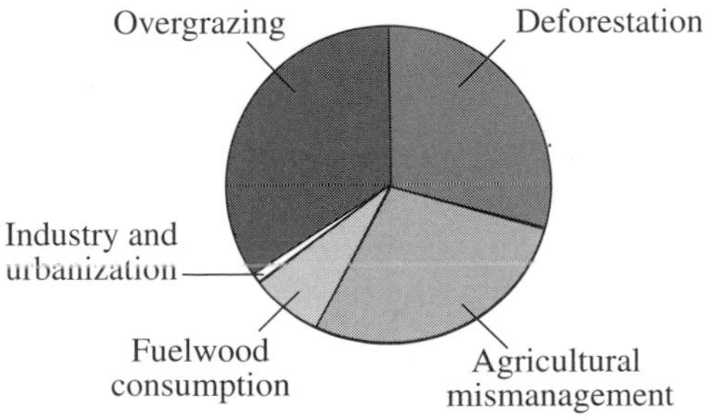

Overgrazing
Deforestation
Industry and urbanization
Fuelwood consumption
Agricultural mismanagement

FIGURE 1-2 • Causes of Land Degradation

TABLE 1-1 Biodiversity of Known Species

	Number of Species	Percent of Total
Animals	1,030,000	72.9
Plants	250,000	17.7
Fungi	69,000	4.9
Protists	58,000	4.1
Monerans	4,800	0.3

Conserving wilderness areas such as rain forests and coral reefs is critical for protecting biodiversity. These natural habitats provide such services to humans as climate control, water purification, and supplies of medicine. To protect biodiversity, environmental activists and scientists hope to increase awareness of the ways in which the activities of humans impact the other organisms that share the planet. The hope is that by increasing such awareness, activities that now threaten the survival of many species will diminish and help slow the rate of extinction.

Refer to Chapters 3 and 4 for more information about conserving natural resources and biodiversity.

CREATING A SUSTAINABLE SOCIAL SYSTEM

The Issues

Poverty is among the most significant contributing factor to environmental degradation, and it is an international problem. More than one billion people, or about one-sixth of the world's population, live under conditions of extreme poverty. Major inequalities between rich and poor exist in many countries. Millions more suffer from malnutrition and disease; many of them live in *developing countries*.

About 25 million people are environmental refugees because they have no home or country of their own. They have been forced to flee their communities or countries because of natural and human-made disasters, such as civil wars, desertification, waste dumps, deforestation, radioactivity, and land loss caused by construction of highways, dams, and other infrastructures. Most of these people become refugees as a the result of incompetent governments which were not prepared to deal with disasters and whose policies destroyed the environment of their communities in the pursuit of profit and progress. The severity of this problem may result in 100 million environmental refugees by the end of 2100 according to some researchers. Additional social problems include *environmental racism*, ethnic persecution, and other human rights abuses.

TABLE 1-2 Income Inequality in Selected Countries, 1990s

Country	Share of Income: Poorest 20 Percent (Percent)	Share of Income: Richest 20 Percent (Percent)
Denmark	9.6	34.5
India	8.1	46.1
United States	5.2	46.4
Russia	4.4	53.7
Zambia	3.3	56.6
Brazil	2.2	64.1

In this table a value of zero indicates perfect equality of income. A value of 100 represents perfect inequality of income and a large gap between rich and poor. The list indicates that Denmark has a better rating of income equality than the other countries on the list. Economists, environmentalists, and social activists are concerned about the widening gap of inequality between rich and poor in many countries.
Source: World Bank.

Environmental Racism

In 1987 the Reverend Benjamin Franklin Chavis, Jr., used the term "environmental racism" to describe the unfair distribution of dumps and incinerators in minority neighborhoods. The term was used in a study conducted by the United Church of Christ Commission for Racial Justice under the leadership of Reverend Chavis. The study defined environmental racism as racial discrimination in environmental policy making, the deliberate targeting of communities of color for toxic waste facilities, and the official sanctioning of the life-threatening presence of poisons and pollutants in minority communities. Concerned that minority populations and low-income populations were bearing a disproportionate amount of adverse health and environmental effects, President William J. Clinton issued an Executive Order in 1994, focusing federal agency attention on these issues. The Environmental Protection Agency (EPA) responded by developing the Environmental Justice Strategy which focuses on addressing these concerns.

Human Rights and a Sustainable Society

Environmental damage is often worse in countries where human rights abuses occur. Where human rights are weak, individuals and community groups are not able to raise environmental issues and solve problems effectively. Individuals need access to justice, the rights to access information, and the freedom of expression to be successful in challenging environmental issues.

Sometimes human rights and environmental abuses are carried out in the name of economic development. Some private companies and corporations expel local peoples from their lands, fail to put into effect pollution-control measures, and subdue local activists who question the impact of their activities on the environment.

Many poor and rural peoples in developing countries are barely informed of projects such as large dams or gas pipelines that claim to provide jobs and reduce poverty but instead displace local inhabitants and exploit their natural resource base. Local communities, including many *indigenous* peoples, suffer environmental abuses when national and state laws do not recognize their rights, particularly to their land and natural resources.

In 1994 an international group of experts on human rights and environmental protection met at the United Nations headquarters in Geneva, Switzerland. They created the Draft Declaration, the first international instrument that links principles of human rights to the environment. Some of the principles in the Draft Declaration include the following:

- All persons have the right to a secure, healthy, and ecologically sound environment.
- All persons shall be free from any form of discrimination in regard to actions and decisions that affect the environment.
- All persons have the right to freedom from pollution, environmental degradation and activities that adversely affect the environment, threaten life, and health.
- All persons have the right to safe and healthy food and water adequate to their well-being.

An indigenous Brazilian rubber plant worker explains how he gathers latex from the seringera trees in Northwestern Brazil. Many indigenous peoples can suffer environmental abuses when national and state laws do not recognize their rights, particularly to their land and natural resources. (Courtesy of Jane Mongillo)

- All persons have the right to a safe and healthy working environment.
- All persons have the right to adequate housing, land tenure and living conditions in a secure, healthy and ecologically sound environment.

In summary, a sustainable environment and economy require social systems that are just and equitable. These strategies include giving local communities, who can initiate a sustainable system, more authority and control over their resources such as access to freshwater and sanitation, better nutrition, job training and education, and the opportunity to create their own wealth.

Refer to Chapters 6 and 7 for more information about community and individual activism.

CREATING A SUSTAINABLE ECONOMY

The Issues

A strong and growing economy has provided a high standard of living for many people; however, economic growth and progress come at a price. Economic activity impacts Earth's natural resources, sometimes with devastating results. The exploitation of natural resources has caused the collapsing of fish stocks, soil erosion, deteriorating rangelands and farms, rising carbon dioxide levels in the atmosphere, mismanagement of forests and croplands, and disappearing wildlife species. Current economic policies require a continuing extraction of natural resources, the production of excessive wastes, and too much dependency on fossil fuels which, critics argue, cannot be sustained. The challenge is to reverse these trends by establishing a new economy—one that emphasizes sustainable development.

A Sustainable Economy

Many argue that economists and ecologists need to work together to design an ecological economy or eco-economy. An ecological economy is based more on using renewable energy sources, such as solar and hydrogen, and less on fossil fuels. An eco-economy relies less on the extraction of minerals and other raw materials, reuses and recycles materials that have already been used, and designs products that last for a long period of time, need little repair, and can be fully recycled. In other words, the ecologically efficient products would be designed to be durable, repairable, and reusable and would be less costly to manufacture. In summary, an eco-economy produces valuable goods while reducing the impact on natural resources.

Some economists and environmentalists would also like to see a change on how a country's economic growth and wealth are measured. Currently, economic growth is measured by the *gross domestic*

TABLE 1-3 Characteristics of a Sustainable Economy

1. Promotes activities that improve the quality of life of all people
2. Uses all natural resources efficiently
3. Relies on renewable technologies
4. Promotes local economies
5. Promotes maximum recycling, source reduction, and reuse
6. Restores damaged ecosystems

product (GDP) of nations. The GDP, a term used to describe a country's economic wealth and economic activity, is the total output of a country's final goods or services valued at market prices, including net exports, during a period of time such as a year.

Environmental economists are critical of the GDP because it does not take into account the losses and depreciation of natural resources used for mining, fishing, logging, farming activities, and industrial uses. The GDP also does not adjust for any loss of natural resources nor does it consider other debits such as air and water pollution. Critics also argue that the GDP includes certain goods and services that make no contribution to the welfare of the people. As a result of these concerns, members attending the UNCED in 1992 proposed measures to use economic and environmental accounting to provide a better tool to measure a country's wealth.

Refer to Chapter 5 for more information about the GDP and economics.

THE SUSTAINABLE DEVELOPMENT MOVEMENT

In the late 1960s, international concerns over environmental problems increased sharply. Many people as well as organizations began to realize that the world's population was growing, and it was putting a great deal of pressure on the natural environment.

Up to that time, environmental problems were handled locally or at the national level. Since deforestation, air and water pollution, and climate changes affect the international community, international efforts are required to solve these environmental problems. This section examines a partial history of the international sustainable movement, beginning in 1971.

Man and the Biosphere Programme

Man and the Biosphere Programme (MAB) an international program established in 1971 by the United Nations Educational, Scientific and Cultural Organization (UNESCO), was set up to provide research and information about natural resources and solutions to environmental issues worldwide. Within the MAB a network of biosphere reserves was established throughout the world. Each biosphere reserve is intended to fulfill three basic functions: conservation, sustainable development, and logistical support for research, education, and information exchange. The program is designed to serve as a basis for rational use and conservation of the resources of the biosphere, and for the improvement of the relationship between people and the environment.

United Nations Environmental Programme

In 1972 the United Nations sponsored the first major international conference on environmental issues held in Stockholm, Sweden. The most important outcome of the conference was the creation of the United Nations Environmental Programme (UNEP).

The UNEP was designed to be "the environmental conscience of the United Nations." It worked to achieve scientific consensus about major environmental issues and to study ways in which to encourage sustainable development in order to increase standards of living without destroying the environment. When the UNEP was created in 1972, only 11 countries had environmental agencies; 10 years later, that number had increased to 106, 70 of them in developing countries.

Convention of International Trade in Endangered Species

In 1975 the Convention of International Trade in Endangered Species (CITES) went into effect with the goal of reducing commerce, the selling and buying, of animals and plants on the edge of extinction. The international agreement prohibits the international trade of any endangered plant or animal species and any products obtained from them such as medicines, furs, and tusks. The agreement also regulates the trade of protected species. The agreement was amended in 1979 to include the Convention on Conservation of Migratory Species of Wild Animals, which protects animals that migrate across borders of different nations. Illegal trade is monitored by Trade Records Analysis of Flora and Fauna in Commerce (TRAFFIC). Today CITES has more than 100 member nations.

Brundtland Report

Although the theory of sustainable development has been around for nearly 30 years, the term became popular in 1987 in the Brundtland Report. The Brundtland Report alerted the world to the urgency of developing an economy that could be sustained without depleting natural resources or harming the environment.

The Brundtland Report, concerned primarily with redistributing resources toward poorer nations while encouraging their economic growth, suggests that equity, growth, and environmental maintenance are simultaneously possible. It also recognizes that achieving equity and sustainable growth would require technological and social changes.

Coalition for Environmentally Responsible Economies

A comprehensive 10-point environmental guideline for corporations, introduced in 1989 as the Valdez Principles, was developed by the Coalition for Environmentally Responsible Economies (CERES). CERES is a nonprofit organization comprising social investors, environmental groups, religious organizations, public pension trustees, and public interest groups. CERES promotes responsible corporate activity for a safe and sustainable future for our planet.

In response to the disastrous Exxon Valdez oil spill in 1989, CERES developed 10 CERES principles to encourage a corporate commitment to a healthy environment. CERES urges all corporations to sign the principles, including

1. Protection of the biosphere
2. Sustainable use of natural resources
3. Reduction and disposal of wastes
4. Wise use of energy
5. Risk reduction
6. Marketing of safe products and services
7. Damage compensation
8. Disclosure of information
9. Employment of environmental directors and managers
10. Annual environmental audits.

By signing the CERES principles, a company makes a pledge to uphold the principles by using only environmentally friendly policies and methods and submitting an annual progress report to CERES and to the public. Although CERES has no enforcement power, it identifies those companies that are not following the principles and urges them to do better.

United Nations Conference on Environment and Development

In 1992 UNCED, also known as the Earth Summit, became the largest gathering of world leaders in history. More than 100 heads of state attended the conference held in Rio de Janeiro, Brazil.

The Earth Summit produced two major treaties. The first was an agreement to reduce the emission of gases leading to *global warming*, and the second was a treaty on biodiversity requiring countries to develop plans to protect endangered species and habitats.

The assembled leaders also adopted Agenda 21, a 300-page plan for achieving sustainable development in the twenty-first century. The attendees adopted Agenda 21 to provide a framework to tackle the worldwide problem of poverty. Countries were urged to cooperate in the essential task of eradicating poverty as an indispensable requirement for sustainable development.

Kyoto Protocol, Earth Summit +5

At the 1992 Earth Summit, the attendees, which agreed on the seriousness of global warming, limited each industrialized nation to emissions in the year 2000 that were equal to or below their 1990 emissions. Unfortunately, these limits were voluntary, and the agreement itself had no enforcement provisions. By 1997 it was clear that these goals would not be met. In a follow-up conference held in Kyoto, Japan, representatives from 160 countries signed a new agreement, known as the Kyoto Protocol, which calls for the industrialized nations to reduce emissions to an average of about 5 percent below 1990 emission levels and to reach this goal by 2012.

At the second conference, held in 1997 in Kyoto, Japan, the parties agreed to reduce greenhouse gas emissions by harnessing the forces of the global marketplace to protect the environment. The Kyoto Protocol's key points, including emissions targets, timetables for industrialized nations, and market-based measures for meeting targets, reflect proposals advanced by the United States. The Kyoto Protocol had to be ratified by at least 55 countries to enter into force.

A central feature of the Kyoto Protocol is a set of binding emissions targets for developed nations. The specific limits vary from country to country, although those for the key industrial powers—the European Union, Japan, and the United States—are similar: 6–8 percent below 1990 emissions levels. The protocol represents only the beginning of securing meaningful developing country participation. Many global warming and climate change researchers believe that the world can

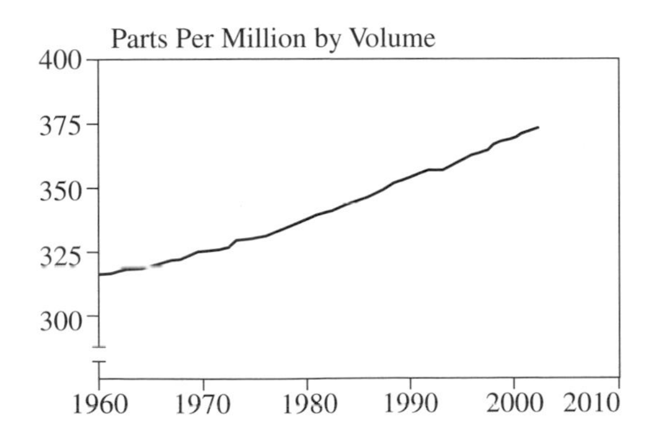

FIGURE 1-3 • Atmospheric Concentrations of Carbon Dioxide, 1960–2002
Source: Scripps Institute of Oceanography

Severn Cullis-Suzuki, an environmental activist, addressed the United Nations Earth Summit held in Rio de Janeiro, Brazil, in 1992. She urged the delegates to work harder on resolving environmental issues. (Courtesy of Jeff Topham)

become more energy efficient over the next 20 to 30 years using current knowledge and today's best technologies. In the longer term, increases in energy efficiency may make it possible to move close to a zero-emissions industrial economy. As of 2004, the United States had not signed the Kyoto agreements.

Refer to Chapter 2 for more information about alternative energies.

World Summit 2002 +10

The Kyoto Summit was followed up by the World Summit on Sustainable Development. The World Summit, held in South Africa in 2002, was attended by more than 65,000 delegates from more than 100 countries. The meeting focused on governments' improving the quality of life for poor countries while reducing environmental degradation and conserving natural resources. Other goals included having countries pledge themselves to slow down biodiversity loss by 2010, restore the collapsing fish stocks by 2015, improve clean water and sanitation access to at least 500 million people by 2015, and improve the manufacturing of toxic materials to safer levels by 2020. Other items on the agenda included greenhouse gas emissions and the Kyoto Protocol, free and fair trade with the elimination of agricultural subsidies used by the developed countries, sustainable energy, corporate responsibility, debt cancellation, and increased foreign aid.

Another task addressed at the World Summit was to assess the goals and pledges stated in the Rio de Janeiro Summit meeting held in 1992. The World Summit 2002 report of those goals was not very favorable. As reported, since 1992, carbon dioxide emissions had increased, coral reefs and forests had continued to decline, and the financial assistance from wealthy nations to developing ones had decreased.

The conclusion of the 2002 World Summit meeting was met with criticism. Although most would agree that the goals were noteworthy, there were no plans of action to implement the goals. One critic remarked that global warming was not a major issue on the World Summit agenda. Another criticism was that although developed and developing countries had agreed to accelerate the substitution of renewable energy for fossil fuels, they failed to demand specific targeted deadlines.

Refer to Chapter 5 for more information on business stewardship.

Individual Stewardship and Activism

According to Mark Hertsgaard, an environmental author, "The history of environmentalism is largely a story of ordinary people pushing for change while governments, corporations, and other established interests

David Brower (1912–2000) often called the Father of Modern Environmental Movement, used his films, advertisements, books, speeches, and other media to promote citizen activism for a sustainable society. Brower founded the Earth Island Institute in San Francisco, California. (Courtesy of www.earthisland.org, Earth Island Institute)

reluctantly follow behind." No question, creating a sustainable future requires individual responsibility and activism. Ordinary people, young and old, must participate in working toward a sustainable society by joining nongovernment organizations (NGOs) and environmental organizations and making changes in their lifestyle habits.

Lifestyle changes include living closer to work; buying recyclable products; purchasing energy-saving light fixtures for the home, school, or business to conserve energy needs; planting trees; and using alternative modes of transportation. Joining an environmental organization can also help one make lifestyle changes as well as participating as a member in resolving environmental issues at the local community, state, and federal levels. Many local communities have organizations that promote recycling programs, car pools, and tips on how to reduce waste.

CAREER CHOICES IN CREATING A SUSTAINABLE SOCIETY

One can also become involved in environmental and social issues by making a career choice. There are many opportunities in the fields of teaching, science, communications, engineering, medicine, business, agriculture, economics, social work, and law studies to identify problems and propose solutions to control pollution, recycle wastes, reduce urban sprawl, implement alternative energy sources, and tackle hunger, disease, and malnutrition problems. The opportunities for individuals to get involved are endless.

Refer to Chapters 6 and 7 for more information about individual activism in environmental issues.

Deep Ecology

Deep ecology is a term coined in 1973 by Norwegian philosopher Arne Naess to describe his belief that humans need to revise their view of nature to recognize its value. To achieve this goal, Naess stressed the need to understand the perspectives of self-realization and biocentric equality. Self-realization requires people to recognize that they are connected with something greater than themselves and to see that the greatest potential for positive life results from its diversity. The idea of biocentric equality, or biocentrism, holds that all organisms have natural value and therefore a right to exist. According to biocentric principles, all species, including those that people may find unattractive, nonuseful, or harmful, should have the same opportunities for protection and survival as species that are considered desirable by human standards. Biocentrism contrasts sharply with anthropocentrism, which measures the value of a species by its usefulness to humans.

Although the term "deep ecology" was not introduced until 1973, naturalist Aldo Leopold is generally called the father of deep ecology. This designation is based on Leopold's essay "A Land Ethic," which appears in his book *A Sand Country Almanac*, published in 1949. In this essay, Leopold stresses the need for people to recognize the intrinsic value of nature and establish a system of values or moral principles about how land is used.

Vocabulary

Agenda 21 The major document that resulted from UNCED, or the Earth Summit, which use held in Rio de Janeiro, Brazil, in 1992. It outlines the extent of global environmental problems and measures needed to ensure sustainable development.

Biodiversity Richness in the number of species—the measure of the variety of Earth's animals, plants, and other organisms and of the ecosystems that support them.

Carrying capacity The maximum number of organisms that can be supported by an ecosystem.

Developing countries Countries that are not fully industrialized.

Economic The management of income, wealth, and expenditures.

Environmental racism The practice of locating a high percentage of hazardous waste sites, incinerators, lead smelter operations, paper mills, chemical plants, and other polluting industries in residential areas of minority communities.

Global warming The predicted excessive warming of the atmosphere resulting from the accumulation of carbon dioxide.

Gross domestic product The total value of a country's annual output of goods and services.

Indigenous Native to a region or country.

Sustainable To keep in existence.

Activities for Students

1. Learn more about Agenda 21, and look at the United Nations' plan for sustainable development. What does it say about children and youth in sustainable development?
2. Research the 50 countries that signed the Kyoto Protocol. What argument would a country make for not signing this document?
3. Start a recycling program for cans, bottles, and papers at your school or in your community.
4. Plan an Earth Day event in your community to inform people about environmental issues.

Books and Other Reading Materials

Brower, David, and Steve Chapple. *Let the Mountains Talk, Let the Rivers Run*. New York: HarperCollins, 1995.

Bullard, Robert D. *Confronting Environmental Racism: Voices from the Grassroots*. Boston: South End Press, 1993.

———. *Unequal Protection: Environmental Justice and Communities of Colorado*. San Francisco: Sierra Club Book, 1994.

Bullard, Robert D., ed. *Dumping in Dixie: Race, Class and Environmental Quality*. 2d ed. Boulder, Colo.: Westview Press, 1994.

Davidson, Eric A. *You Can't Eat GNP: Economics as if Ecology Mattered*. Cambridge, Mass.: Perseus Publishing, 2000.

Hawken, Paul. *The Ecology of Commerce, a Declaration of Commerce*. New York: Harper Business, a division of HarperCollins, 1993.

Hertsgaard, Mark. *Earth Odyssey: Around the World in Search of Our Environmental Future*. New York: Broadway Books, a division of Random House, 1998.

Johnson, Huey D. *Green Plans, Greenprint for Sustainability*. Lincoln: University of Nebraska Press, 1997.

Lerner, Steve. *Eco-Pioneers*. Cambridge, Mass.: Massachusetts Institute of Technology Press, 1997.

Mongillo, John, and Bibi Booth. *Environmental Activists*. Westport, Conn.: Greenwood Press, 2001.

Mongillo, John, and Linda Zierdt-Warshaw. *Encyclopedia of Environmental Science*. Phoenix, Ariz.: Oryx Press, 2000.

Schwartz, Linda, and Beverly Armstrong. *Earth Book for Kids: Activities to Help Heal the Environment*. Huntington Beach, Calif.: Learning Works, a Division of the Creative Teaching Press, 1990.

Sitarz, Daniel. *Agenda 21: The Earth Summit Strategy to Save Our Planet*. Boulder, Colo.: Earth Press, 1993.

Websites

Agenda 21, since its inception and about the Earth Summit, http://www.un.org/esa/earthsummit/index.html or http://www.un.org/dpcsd/ earthsummit

A Solutions Site for Kids, http://www.soultion-site.org/kids/

CITES homepage, http://www.cites.org

Earth Day Network, worldwide@earthday.net, http://www.earthday.net/

Earth Times, leading independent, international, nonpartisan newspaper focusing on environment and sustainable development, http://www.earthtimes.org

Energy Efficiency and Renewable Energy Network, www.sustainable.doe.gov/

Environmental News Network, http://www.enn.com/

North American Association for Environmental Education, http://naaee.org

Global Response, empower people of all ages, cultures, and nationalities to protect the environment by creating partnerships for effective citizen action, http://www.globalresponse.org/

Poverty, http://www.doc.mmu.ac.uk/aric/eae/english.html

Turn the Tide, http://www.newdream.org/turnthetide/

Virtualhouse, virtual tour of a house to learn about our connections to biodiversity, http://www.virtualhouse.org

CHAPTER 2

Sustainable Energy Systems

Energy is the lifeblood of the global economy. Energy, in the form of electricity, is used to light and heat homes, industries, farms, and businesses. Chemical energy, in the form of fuels, is used to run cars, buses, trains, airplanes, and other machines. Energy is vital for economic growth and development. Much of the energy used today in the global industrial and agricultural economy comes from fossil fuels, which are fundamental natural resources. Currently, it is estimated that much of the energy consumed in the United States and the world comes from fossil fuels.

Refer to Volume II for more information about energy and energy systems.

IMPACT OF FOSSIL FUEL EMISSIONS ON THE ENVIRONMENT

Large demands for fossil fuels began in the twentieth century as a result of the invention of gasoline- and diesel-powered vehicles, as well as the spread of technology and new inventions. By 2001 approximately 90 percent of the globe's commercial energy use was derived from the consumption of coal, petroleum, and natural gas. When large volumes of these fuels are burned, however, *carbon dioxide* and other greenhouse gases are released into the air causing serious environmental pollution problems such as smog, acid rain, and global warming.

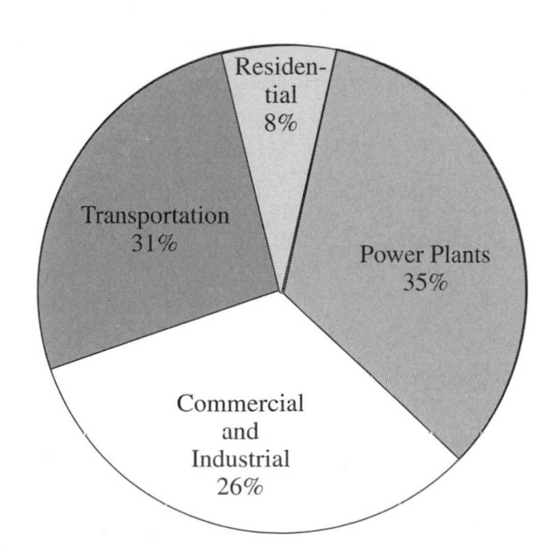

FIGURE 2-1 • **Sources of Carbon Dioxide**

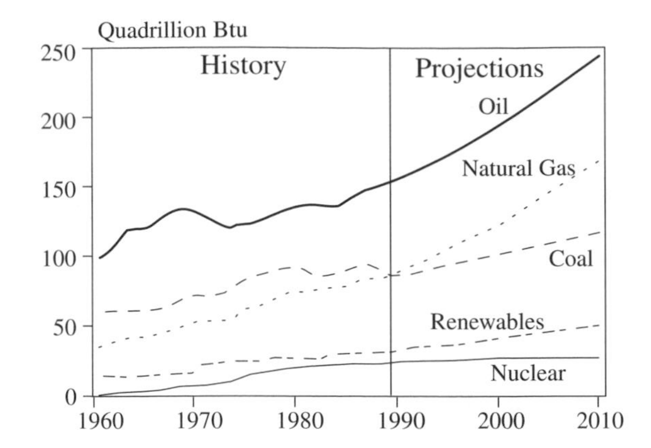

FIGURE 2-2 • World Energy Consumption by Fuel Type, 1960–2010
Source: Energy Information Administration/International Energy Outlook, 2001

Refer to Volume II for more information about fossil fuels.

Fossil fuel emissions are responsible for about 80 percent of the world's carbon dioxide emissions and other harmful byproducts.

CHANGING TO SUSTAINABLE ENERGY SYSTEMS

Today, government agencies, businesses, farmers, major energy corporations, environmental groups, and economists are joining together to develop and provide a more sustainable energy system for the future—an alternative plan to fossil fuel energy. A sustainable energy system includes *renewable energy* resources such as wind power, solar energy, hydrogen and fuel cells, geothermal energy, biomass, and hydroelectric power.

Although the technology for expanding the use of alternative energy sources is growing, these resources still do not supply very much of the world's energy. Approximately 60 percent of the world's electricity and about 80 percent of its nonelectrical energy needs are produced by fossil fuels. By the middle of the twenty-first century, economic analysts and environmentalists predict that renewable energy sources, particularly wind energy, direct sunlight, and hydrogen fuel, could supply about 60 percent of the world's electricity and about 40 of its nonelectrical energy requirements.

SUSTAINABLE ENERGY SYSTEMS

Wind Energy

The fastest growing renewable power source is wind energy. Wind energy, or wind power, is an alternative energy resource that uses the renewable energy in moving air to generate electricity. Although wind power currently produces less than 2 percent of the world's electricity, the Worldwatch Institute estimates that wind energy could easily provide between 20 and 30 percent of the electricity needed by many countries. In the United States, the American Wind Energy Association (AWEA) estimates that, by the year 2025, wind power will produce more than 10 percent of the electricity in the United States.

Wind energy is the *kinetic energy* associated with the movement of atmospheric air. Wind energy systems convert this kinetic energy to more useful forms of power. Wind turbines, or aerogenerators, are used to generate electricity from wind. Most often, wind turbines are installed in large numbers in wind farms. Sites suitable for use as wind farms are usually located in areas that regularly receive sustained winds

The first large-scaled Native American owned and operated wind farm is located on Rosebud Sioux reservation in south-central South Dakota. (Courtesy of Bob Gough, Intertribal COUP)

In the United States, wind farms are being developed in many sections of the west and mid-west. This wind farm is in Tehachapi, California. (Courtesy of Warren Gretz/NREL, National Renewable Energy Laboratory)

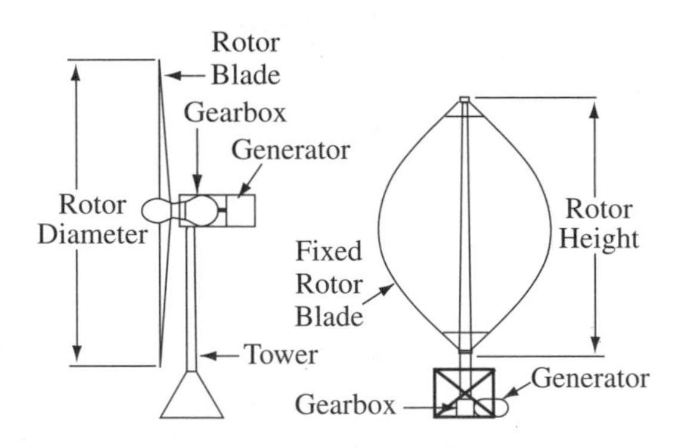

FIGURE 2-3 • Wind Turbine Design

of at least 22.5 kilometers per hour (14 miles per hour) and are not blocked by obstacles, such as high mountains.

According to the Earth Policy Institute, advances in wind turbine technology have lowered the cost of wind power from 38 cents in the early 1980s to from 3 to 6 cents today. As a result of the lower costs, pollution-free energy, and better turbine technology, the United States and other countries are aggressively building wind farms.

U.S. WIND FARMS

In 2001 a 300-*megawatt* wind farm was constructed on the Oregon/ Washington border; as of that date, it was the world's largest wind farm. The wind farm is projected to supply electricity to 70,000 homes and businesses. A 3,000-megawatt wind farm is being built in South Dakota on the Iowa border, and when it is completed it will be 10 times the size of the Oregon/Washington wind farm. Wind farms are being con-

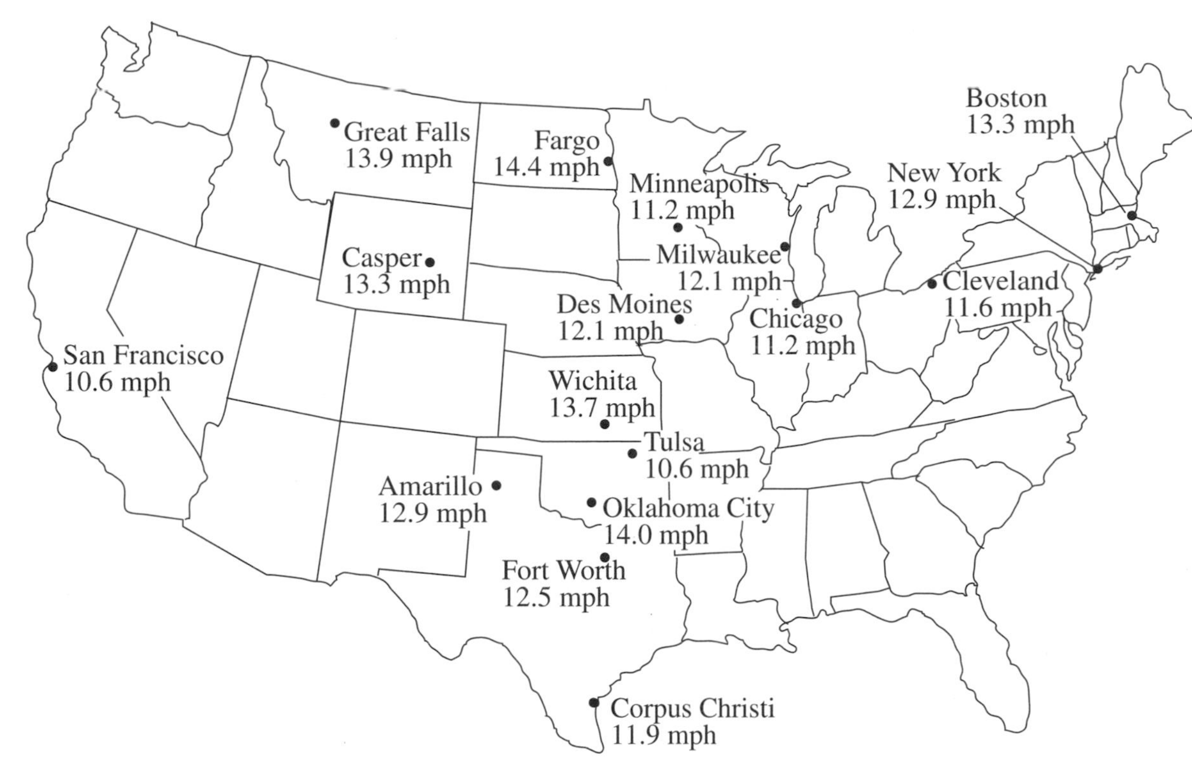

FIGURE 2-4 • The map shows wind conditions in some areas of the United States that may have the potential for wind power energy developments.

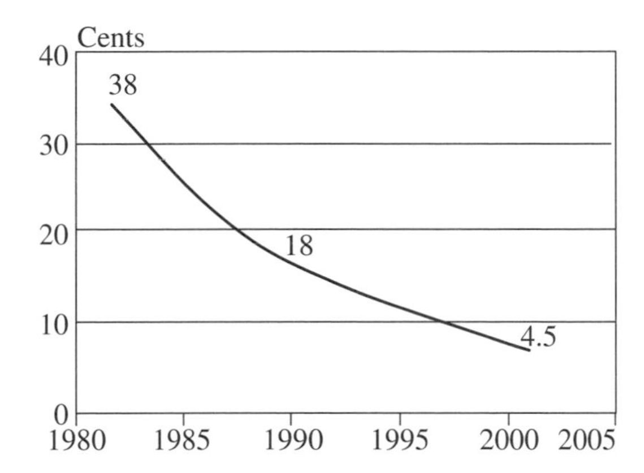

FIGURE 2-5 • Average Cost per Kilowatt-hour of Wind-powered Electricity in the United States, 1982, 1990, and 2001

Source: Eco-Economy, by Lester R. Brown, p. 105. American Wind Energy Association.

structed in Minnesota, Iowa, Texas, Colorado, Wyoming, and Pennsylvania. California was one of the first states to develop wind farms. Altamont Pass, one of the state's wind farms, is located east of San Francisco among a series of low hills which separate the San Francisco Bay area from the San Joaquin Valley. Another California wind farm, called Tehachapi Pass, is located in the Mojave Desert north of Los Angeles. Its 5,000 wind turbines generate enough electricity to meet the residential needs of thousands of southern Californians, making the site one of the world's largest producer of wind-generated electricity. In all, 26 states in the United States produce wind energy.

DID YOU KNOW?

According to the American Wind Energy Association, the wind conditions in Texas, North Dakota, South Dakota, and Kansas can generate enough wind energy for the whole nation.

WIND FARMS IN OTHER COUNTRIES

The leading countries that account for 80 percent of the world's installed wind energy capacity include Denmark, Germany, the United States, Spain, and India. Other countries, including Great Britain, France, Argentina, Brazil, and China, are now working on plans to increase their use of wind power.

The Danes were the first people regularly to produce electricity using wind power; by World War I, Denmark had a network of wind turbines which generated about 100,000 kilowatts of electricity. Today, Denmark gets 18 percent of its electricity from wind power. Denmark is also the world leader in wind turbine technology and manufacturing. Danish-built wind turbines provide 55 percent of the world's wind power and rank third in total wind-generating capacity. The country is now building the world's largest offshore wind farm approximately 32 kilometers (20 miles) off the shore of Denmark.

The use of wind power is growing most rapidly in Germany. In 1994 the recently reunified country surpassed Denmark as the world's second largest wind energy powerhouse. Wind farms in Germany are concentrated along the coastal areas in the northwestern portion of the country. One of the northernmost German states gets 19 percent of its electricity from wind energy.

The development of wind power as a source of electricity in Great Britain began in the early 1990s, when 10 wind turbines were installed on a farm in Cornwall. Today, most of the United Kingdom's wind plants are located in England and Wales, and more wind farms are scheduled to be built in Scotland.

Spain is moving up rapidly in the European wind power market place. The state of Navarra, in Spain, receives 24 percent of its electricity from wind. France has announced that it has plans to develop 5,000 megawatts of wind power by 2010. Today Europe accounts for 70 percent of the world's wind power.

DID YOU KNOW?

The average house uses between 6,000 and 7,500 kilowatts of electricity per year.

Refer to Volume II for more information about wind energy.

The South American countries of Argentina and Brazil are developing wind power technology as well. The goal in Brazil is to install wind turbines that will produce 1,000 megawatts of electricity by the year 2005. Argentina's goal is to produce 3,000 megawatts of electricity by 2010.

In Asia, India's wind power is now ranked fifth in the world. In fact, India is also the world's fifth largest generator of solar power. China is planning to develop 2,500 megawatts of wind power by 2005.

Solar Energy

Solar energy is another sustainable energy system. Solar energy is the conversion of direct radiant energy from the sun into other forms of energy to provide heat and electricity. Different solar technologies include passive systems, active systems, and solar cells. The passive solar

Seabed Ocean Turbines

Windmill-like underwater turbines can be used to harness ocean tides to produce electricity. The world's most northerly town, Kvalsund, in the Arctic area of Norway, is preparing to receive its electricity from an underwater turbine power station which will be built on a nearby seabed. The underwater tidal turbines, which weigh about 200 tons, are installed deep below the keels of passing ships to prevent accidents.

A seabed turbine is designed to move always in the direction of incoming or outgoing tidal currents, even when the currents change direction. The movement of the seawater turns the blades of the turbine to generate electrical power. The quiet seabed turbines have slow-moving blades which pose negligible danger to marine life but allow water and silt to flow freely. This feature allows fish to swim around the turbine without being harmed.

The turbines are designed to be maintenance free for three years, then divers can access them if needed to perform repair work. When in full production, the Norwegian turbine will generate electricity for approximately 1,000 homes.

Proponents of seabed turbine energy believe there are many advantages of using underwater tidal turbines. The advantages include low operating costs, zero greenhouse gas emissions, and low ecological impact. Furthermore, seabed turbines can provide more power than wind or solar technologies in a comparable area.

A model of a wave power plant that can use underwater tidal turbines to generate electricity. (Courtesy of Energetech Australia Pty Limited)

heating system, which relies largely on the greenhouse effect, traps heat inside a building much as a closed automobile traps heat when parked in an unobstructed area on a sunny day. Active systems use special rooftop panels that collect sunlight. The fluid in the panels absorbs the heat, which can be stored for a variety of uses, particularly hot water.

The solar cell, or *photovoltaic* (PV) cell, is a device that converts solar energy into electricity in a manner that does not release any pollutants into the environment. According to research, about 40 percent of all solar cells sold are for producing electricity for homes and for pumping water, and about 35 percent of them are used to power transmitting and communication operations. Common solar cells are used to run calculators and watches.

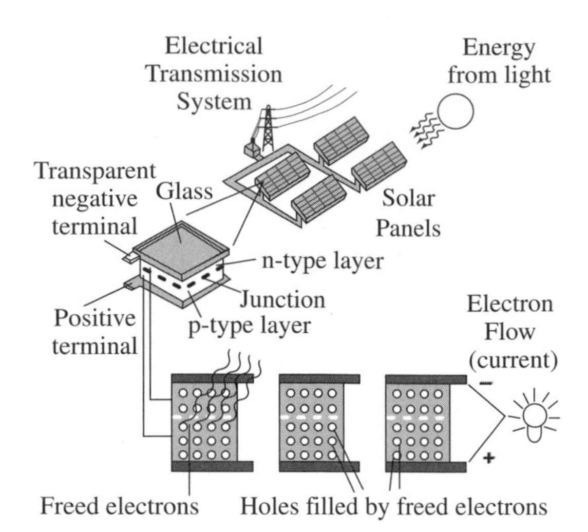

FIGURE 2-6 • Inside a Photovoltaic Cell

HOW SOLAR CELLS WORK

The process of producing electricity from a solar cell begins with sunlight. Sunlight contains energy in the form of photons or particles of light. When photons strike a solar cell, they cause electrons to be ejected from the silicon atoms located near the junction. The stream of electrons moves freely from the negative layer to the positive layer through the metal terminals. A 10-centimeter (4-inch) solar cell can produce about 1 watt of direct current (dc) electricity when exposed to sunlight. To generate more electricity, many solar cells are wired together in a panel called a solar array, which is encased in a watertight container. The panels, in turn, can be wired together to generate an even greater amount of electricity. These types of solar panels are placed on the roofs of homes and businesses to generate electricity.

SOLAR ENERGY USES

In 2003 Japan and Germany were the world's largest generators of solar energy power. In Japan 80,000 households have installed solar roof panels to provide their electric needs. In 1998 Germany began its own program to equip 100,000 homes with solar rooftops. South Africa is planning the installation of 350,000 solar homes in rural areas. More than 500,000 homes throughout the world use solar cells as a source of electricity.

Solar cells can also produce electric power in remote rural communities where it would be difficult and costly to install and connect electric power lines. The Northwest Rural Public Power District in South Dakota, along with many other utilities around the country, offers customers in remote areas the option of installing a solar cell system. On some South Dakota farms, solar cell–powered water pumps are particularly popular in rural areas for irrigation and other water uses.

More than a million rural homes in developing countries receive electricity from solar cells. For example, in the remote village of

According to some experts, a solar panel measuring 5 meters square could generate 600 watts an hour, which is enough for a typical home. The cost for the solar panel would range between $15,000 to $20,000 at current market prices. However, these estimates vary considerably depending on where the house is located and how much electricity a family needs. (Courtesy of National Renewable Energy Laboratory)

Cacimbos in the state of Ceará, Brazil, each home has a small 50-watt solar cell system which provides some electricity for lights. The low costs of solar cells have made them more affordable to South American countries and other countries such as India, China, South America, and Vietnam, where people lack electricity in rural or remote areas. Solar cells are also used in lighthouses, on offshore petroleum drilling operations, and in radio and telephone transmitters.

Although solar energy costs are higher than fossil fuel and hydroelectric power, they continue to decline each year. According to one report, in 1992, solar energy costs ranged between 40 and 50 cents per kilowatt-hour, or about five times more than fossil fuel energy costs. In 2002, however, solar energy costs were only about two times more expensive than fossil fuel. Solar energy costs continue to decline and become more affordable.

DID YOU KNOW?

In 1839 French physicist Edmond Becquerel observed that, when certain materials absorbed light, the materials generated electricity. Becquerel also recorded that the amount of electricity varied with the intensity of the light. Despite these early findings by Becquerel, photovoltaic research did not begin in earnest until the late-1950s. NASA first used photovoltaics in 1958 to power the radio of the U.S. *Vanguard I* space satellite with less than 1 watt of electricity.

Geothermal Energy

Natural heat energy obtained from the interior of Earth, which is extracted from sources such as steam, hot water, and hot dry rocks, is known as geothermal energy. Geothermal energy, a renewable energy resource, can be used for the direct heating of buildings or for generating electricity. In some cases, however, geothermal energy is not listed as a renewable energy source because the depletion rate of sources such as hot water can be higher than the rate of replenishing or recharging the sources.

Geothermal energy is used widely in many parts of the world, including Italy, Iceland, New Zealand, Russia, Japan, France, the United States, the Philippines, Indonesia, Mexico, Kenya, El Salvador, and Nicaragua.

The Geysers Geothermal Field, located in northern California, is the largest geothermal electric power plant in the world. It produces about 1,300 megawatts of electricity, enough to satisfy the residential electricity needs of 1.7 million Californians—more than the combined

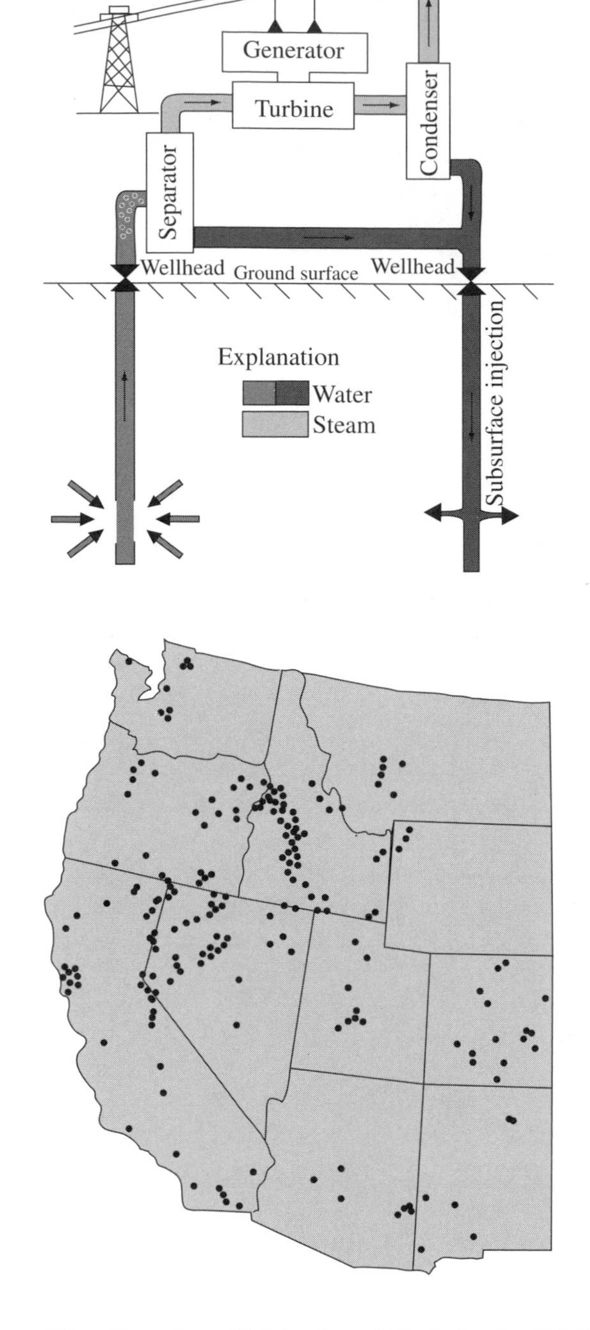

FIGURE 2-7 • A Hot Water Hydrothermal System The diagram illustrates how electricity is generated from a hot-water hydrothermal system. The part of hydrothermal water that becomes steam is separated and used to drive a turbine generator. Wastewater from the separator and condenser is injected back into the subsurface to help replace the groundwater used in the hydrothermal system. *Source:* Geothermal Education Office and the United States Geological Survey.

FIGURE 2-8 • Major Geothermal Sites in the United States

populations of San Francisco, Oakland, and Berkeley in California's Bay Area. According to the U.S. Energy Information Agency, geothermal energy has the potential to provide the United States with 12,000 megawatts of electricity by the year 2010 and 49,000 megawatts by 2030. Other U.S. geothermal sites are located in Hawaii, Nevada, Wyoming, and Utah.

In Iceland, 65 percent of the homes are heated by geothermal energy; in El Salvador, geothermal energy generates about 40 percent

California Geothermal Geyser Field. (Courtesy of National Renewable Energy Laboratory)

of the country's electricity needs; in Kenya, about 11 percent of the country's electricity; and in Nicaragua, about 28 percent of the country's electricity.

Geothermal energy has proven to be extremely reliable and flexible. One of the most important advantages of geothermal energy is that it does not pollute the atmosphere with sulfur dioxide and nitrogen emissions as other power plants using fossil fuels do. Geothermal energy also has some disadvantages. The construction of geothermal plants in rain forests can destroy sensitive ecological habitats, and the drilling of wells can disrupt underground faults and fissures, which may lead to seismic activity and landslides.

Refer to Volume II to learn more about the geothermal process.

DID YOU KNOW?

Geothermal heat sources were first used for electrical power production in Italy in the year 1903. The plant is located at the Larderello geothermal field, and electrical power is still being produced there.

Biomass

Biomass supplies 20 percent of the world's energy mostly as a source for heating and cooking, rather than electricity. Biomass includes such material as wood, herbaceous plants, excess food crop, animal wastes,

The soybean-powered bus burns cleaner than diesel with fewer emissions. Soybean fuel is safe to handle, non-toxic, biodegradable, and requires no engine modifications or special fueling facilities like other alternative fuels. (Courtesy of U.S. Department of Energy)

and municipal solid wastes. Firewood is the best known, most widely used biomass fuel. More than 1 billion people use firewood as their main energy source for cooking and heating.

Biomass can be burned directly as a solid fuel, or it can be used to produce liquid and gas fuels, known as biofuels. Two liquid forms of biofuels include methanol, a wood alcohol, and ethanol, a grain alcohol. Methanol can be produced from fossil fuels such as natural gas and coal as well as from the fermentation of grains from agricultural, forestry, and municipal wastes. Ethanol is produced primarily from the fermentation of grains such as corn and wheat and from plants containing starches and sugars, such as sugarcane. Both methanol and ethanol fuels are combined with gasoline. Methanol (M85) is 85 percent methanol and 15 percent gasoline, and ethanol (E85) is 85 percent ethanol and 15 percent gasoline. A car designed to use ethanol or methanol can be from 10 to 20 percent more efficient than a fossil fuel car. About 5 percent of the energy consumed in the United States is provided by biofuels.

In France, Italy, and Germany, biofuels for diesel engines are produced from domestic oilseeds and cottonseeds and even from vegetable oils and animal fats. Another source of biomass energy is methane, a biogas that can be produced from manure and from landfill waste sites. Biofuels are cleaner than fossil fuels because they release few greenhouse gases, such as carbon dioxide and sulfur, and particular matter into the atmosphere.

Hydroelectric Power

Hydroelectric power accounts for about 20 percent of the world's electricity. Hydroelectric power, or hydropower, is an alternative energy resource which has been used since the nineteenth century. It uses flowing water to drive turbines to generate electricity.

Some of the largest hydroelectric power producers are Canada, the United States, Brazil, Norway, Russia, and China. About 10 percent of all U.S. electricity is produced by hydroelectric power. In Canada, hydroelectric power provides 12 percent of the country's total annual energy demand. By contrast, hydroelectric power accounts for less than 2 percent of the electricity produced in the United Kingdom.

DID YOU KNOW?

The world's first hydroelectric power station went into operation on the Fox River in Appleton, Wisconsin, in 1882. It generated a total of 25 kilowatts of power.

The world's largest hydroelectric plant, as of 2003, is the Itaipú hydroelectric power plant, located on the border between Brazil and Paraguay. Under construction from 1975 to 1991, Itaipú represents the efforts and accomplishments of two neighboring countries, Brazil and Paraguay. In 1995 Itaipú alone provided 25 percent of the energy supply in Brazil and 78 percent in Paraguay.

A major benefit of hydroelectric power is that it is nonpolluting: it produces no harmful emissions such as carbon dioxide (CO_2), sulfur dioxides (SO_2), or nitrogen oxides (NO_x) and no liquid or solid wastes. However, there are concerns about and opposition to building large dams for hydroelectric power because they negatively affect the environment.

FIGURE 2-9 • The Itaipú Hydroelectric Power Plant is the largest of its kind in the world. Its location on the Paraná River allows Brazil and Paraguay to share its output. In 1995, Itaipú alone provided 25 percent of the energy supply in Brazil and 78 percent in Paraguay. The Itaipú hydroelectric power plant generates more electrical energy than any other dam in the world.

The Hoover Dam provides hydroelectric power for use in Nevada, Arizona, and California. Hoover Dam alone generates more than 4 billion kilowatt-hours a year—enough to serve 1.3 million people. (Courtesy of U.S. Department of Interior. Bureau of Reclamation, Andrew Pernick, Photographer)

Building small, rather than large, hydroelectric power systems may be the trend for the future as in the past. Today, small-scale hydroelectric power systems, called "mini-hydro" or "micro-hydro" systems, are being used on rivers and tributaries and in remote areas. Such small-scale systems may not require the damming of rivers. These mini-hydro systems are used in China and the United States, as well as in Indonesia, Nepal, Sri Lanka, and Zaire. Small-scale hydropower can be used locally in remote villages and towns to generate electricity for businesses, farming, and lighting, as well as for pumping water.

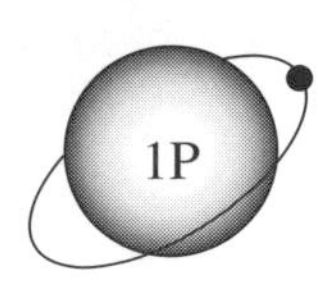

FIGURE 2-10 • Much of the current industrial uses of hydrogen is in the production of ammonia (about 50 percent). About 36 percent of hydrogen production is used in petroleum refining and the rest is used for methanol production and other uses.

Hydrogen and Fuel Cells

Many engineers, businesses, environmentalists, and government organizations propose that fuel cells and hydrogen fuel will be the cornerstone of sustainable development. In fact, many experts predict that a hydrogen fuel cell economy will take place later in this decade as conventional nonrenewable resources become less available.

HYDROGEN AND MAKING HYDROGEN

Hydrogen, a colorless, odorless gas existing as an H_2 molecule, makes up about 75 percent of the universe's mass. Hydrogen is found on Earth only in combination with other elements such as oxygen, carbon, and nitrogen. Water, all organic matter, and acids contain hydrogen molecules.

Since hydrogen does not occur freely on Earth, breaking it free from other compounds is difficult. However, hydrogen can be separated from various fossil fuels such as coal, methane, natural gas, and petroleum. In fact, hundreds of tons of hydrogen are generated daily from natural gas at refineries. Hydrogen can also be separated from water by using electricity produced from fossil fuels or from renewable sources such as solar energy, wind, or geothermal energy.

DID YOU KNOW?

Hydrogen forms about 11 percent by weight of water (H_2O) and is not very reactive at room temperatures. At high temperatures, however, it burns vigorously and often explosively.

Today, hydrogen is used primarily in ammonia manufacturing, a major component of fertilizer, and in petroleum refineries to remove sulfur from gasoline. Food manufacturers use some hydrogen, in a process called hydrogenation, to convert oils into margarine.

HYDROGEN FUEL CELLS

Hydrogen fuel cells are devices that directly convert hydrogen into electricity. Fuel cells can provide electricity to power motor vehicles and to heat and light homes, office buildings, and factories. Fuel cells are used in NASA's space program to provide heat, electricity, and drinking water for astronauts. Fuel cells are also used aboard Russian space vehicles.

Different types of fuel cells use different *electrolytes*, operate at different temperatures, and are suited to different uses. Currently, the most popular fuel cell is called the proton exchange membrane (PEM). A lightweight fuel cell, it is one of the easiest to build. The outside portion of the fuel cell, or membrane, is coated with platinum which acts as a *catalyst*. Hydrogen under great pressure and temperature is forced through the catalyst. At this point, the element is stripped of its electrons, allowing them to move through a circuit to produce electricity.

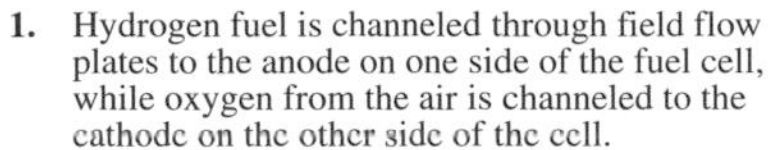

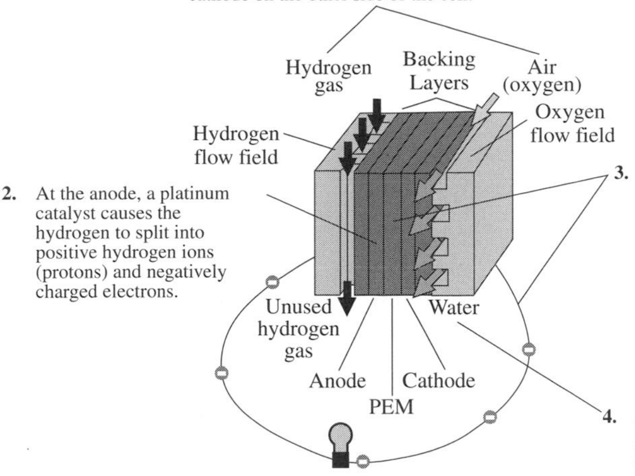

FIGURE 2-11 • Fuel Cell Source Future vehicles and electric generators powered by fuel cells could eventually run on hydrogen derived from corn. The task will be to convert ethanol produced by the fermentation of corn into hydrogen fuel. *Source:* U.S. Department of Energy

The hydrogen protons pass through the membrane and combine with oxygen in the outside air to form water.

The hydrogen used in most fuel cells is made from "reforming" methane with high-pressure steam. The steam interacts with the methane to separate the hydrogen from the *hydrocarbon* molecules in the methane. Special equipment, called reformers, can make hydrogen from coal, fossil fuels, nuclear energy, sewage, bacteria, and paper-mill waste.

FUEL CELLS FOR AUTOMOBILES

The next technological advance in automobile power will be fuel cells. The hydrogen fuel cell vehicle is an excellent alternative to fossil fuel vehicles because hydrogen produces no carbon dioxide when burned and because the fuel cell requires little maintenance because it has so few moving parts. Within the next decade, many automakers and other experts believe the hydrogen fuel cell will replace the need for petroleum, diesel, and natural gas as the main fuel for automobiles, buses, and trucks. Carmakers have invested billions in fuel cell research and are currently testing prototype fuel cell vehicles; however, the mass production of fuel cell vehicles for consumers will not take place until 2010.

DID YOU KNOW?

According to some fuel cell experts, fuel cells, stacked up to the size of a tall bookcase, could provide all the electrical needs of a standard home without using power lines.

For now, the alternatives to fossil fuel, internal combustion engines are electric and hybrid vehicles. In 1996 General Motors introduced battery-powered electric vehicles in the United States. Electric vehicles (EVs) have also been built by automakers in Japan and other countries in Asia, as well as in France, Switzerland, Germany, and Italy. Electric vehicles have stopped being popular because the heavy-built lead-acid batteries installed in the vehicles need recharging after short trips, and the life of the battery is about three years.

On the other hand, the hybrid electric vehicle uses two sources of power: an electric motor powered by batteries and a gasoline engine. The gasoline engine provides some power, and is used as a generator to recharge the batteries. Toyota was the first automaker to start selling a full-production hybrid vehicle, the Prius, which went on sale in Japan in 1997. The car is also sold in the United States. The early Toyota hybrid car, which used a nickel-metal hydride battery, achieved speeds of 160 kilometers (100 miles) per hour when the engine and electric motor were used in combination. Other carmakers, such as Honda, have built and sold hybrids.

There are environmental concerns with the use of batteries for EVs and hybrid cars. Although EVs produce zero emissions of their own, the electricity used to charge EVs can originate from fossil fuel, electric power plants which produce airborne pollutants and solid waste at the plants. Batteries also contain toxic chemicals and produce some toxic emissions which make battery production and the disposal of worn-out batteries a waste issue. Other criticisms include the assertion that today's batteries are still too expensive, store too little energy, are too heavy, and need to be replaced eventually.

Fuel cells will make electric cars more practical. Unlike a battery, a fuel cell does not run down or require recharging; it generates energy as long as a fuel is supplied. In addition, fuel cells are quiet and flexible, operate at low temperatures, and produce lower emissions than gasoline-powered models.

Several auto companies are currently working on fuel cell technology. In 1999 the DaimlerChrysler automobile company introduced a fuel cell car in the United States. The electric-powered car, the NECAR 4, was chosen at the 1999 International Engine of the Year Awards as the best engine concept for the future. According to the automaker, NECAR 4, based on the Mercedes-Benz car, operates

Chrysler Fuel Cell Hybrid Vehicle. In 2003 and 2004, a fleet of sixty of these vehicles are being tested in Europe, the USA, Japan and Singapore. (Courtesy of DaimlerChrysler)

Toyota Fuel Cell Hybrid Vehicle. For Toyota's environmental technologies pages: http://www.toyota.com/about/environment/technology/fchv.html. (Courtesy of Warren Gretz/NREL, National Renewable Energy Laboratory)

on liquid hydrogen in the fuel cells which generate electrical energy to power the vehicle. NECAR 4, a zero-emission vehicle, can travel about 448 kilometers (280 miles) before the hydrogen used in the fuel cell needs to be replenished. The car reaches a top speed of 144 kilometers (90 miles) per hour and provides ample room for five occupants and their luggage.

Toyota has also been working on hydrogen-fueled vehicles. Since 1992, Toyota has been developing fuel cell vehicles. In 2003 the FCHV-4 and FCHV-5 (fuel cell hybrid vehicle) were the company's latest prototypes. Built on a Toyota Highlander body frame, the FCHV-4 is currently undergoing on-road testing in both Japan and America. Other companies working on fuel cell vehicles include Ford and General Motors. Most carmakers expect that fuel cell vehicles will need more time for development and will not be mass-produced until 2010. The future of fuel cells, however, looks very bright.

DID YOU KNOW?

The first fuel cell was built in 1839 by Welsh judge and scientist Sir William Grove. The use of fuel cells as a practical generator began in the 1960s, when NASA chose fuel cells over nuclear power and solar energy to power the Gemini and Apollo spacecraft. Today, fuel cells provide electricity and water on the space shuttles.

Vocabulary

Carbon dioxide Gas produced when carbon combines with oxygen and has no characteristic color or odor.

Catalyst A substance which will change the speed of a reaction without being itself permanently changed. Water is a catalyst for many reactions, such as the rusting of iron.

Electrolyte A substance which will conduct a current when in a solution or melted form.

Hydrocarbon A chemical compound containing carbon and hydrogen as the principal elements.

Kinetic energy Energy of motion.

Megawatts 1,000,000 watts.

Photovoltaic Solid-state silicon electrical devices which convert sunlight directly into an electric current.

Renewable energy Power from any source that replenishes itself.

Watt A unit of power equal to one joule per second.

Activities for Students

1. Iceland uses geothermal sources for 65 percentage of its energy production. Investigate which geographical features are found in Iceland to make geothermal energy a practical option there.
2. What are some of the environmental consequences of using hydroelectric power? Research how dams disrupt natural habitats of plants and animals.
3. Create a pro and con list of whether incentives should be offered to drivers of hybrid or hydrogen automobiles. Consider who would provide the incentives, and whether provisions should be made in state or local laws to encourage the use of alternative fuel vehicles.

Books and Other Reading Materials

Ausubel, Ken. *Restoring the Earth: Visionary Solutions from the Bioneers*. Tiburon, Calif.: Kramer, 1997.

Brown, Lester. *Eco-Economy: Building an Economy for the Earth*. New York: W. W. Norton, 2001.

Duffield, W. A., J. H. Sass, and M. L. Sorey. *Tapping the Earth's Natural Heat*. U.S. Geological Survey Circular 1125, 1994. A full-color book available at: http://pubs. usgs.gov/2004/c1249/ that describes, in non-technical terms, USGS studies of geothermal resources-one of the benefits of plate tectonics-as a sustainable and relatively nonpolluting energy source. U.S. Geological Survey, Branch of Information Services, P.O. Box 25286, Denver, CO 80225.

Graham, Ian. *Geothermal and Bio-Energy (Energy Forever)*. Austin, Texas: Raintree/Steck Vaugh, 1999.

Hoffmann, Peter. *Tomorrow's Energy: Hydrogen, Fuel Cells, and the Prospects for a Cleaner Planet*. Cambridge, Mass.: MIT Press, 2001.

Houghton, John Theodore. *Global Warming: The Complete Briefing*. Cambridge, England: Cambridge University Press, 1997.

Kordesch, Gunter, and Karl Kordesch. *Fuel Cells and Their Applications*. New York: John Wiley, 1996.

Websites

American Wind Energy Association, http://www.igc.apc.org/awea/news/html

Center for Renewable Energy and Sustainable Technology (CREST), Solar Energy Research and Education Foundation, http://solstice.crest.org/

Electric Vehicle Association of the Americas, http://www.evaa.org

Electric Vehicle Technology, http://www.avere.org/

Encyclopedia of the Atmospheric Environment, http://www.doc.mmu.ac.uk/aric/eae/

Geothermal Database USA and Worldwide, http://www.geothermal org

Office of Energy Efficiency and Renewable Energy (part of U.S. Department of Energy), http://www.eren.doe.gov/

Solar Energy Industries Association, http://www.seia.org/main.htm

U.S. Bureau of Reclamation, Hydropower Information, http://www.usbr.gov/power/edu/edu.htm

U.S. Department of Energy, http://www.doe.gov

U.S. Department of Energy, Alternative Fuels Data Center, http://www.afdc.nrel.gov

U.S. Department of Energy, Office of Fossil Energy, http://www.fe.doe.gov

U.S. Department of Energy, Photovoltaic Program, http://www.eren.doe.gov/pv/text_frameset.html

U.S. Geological Survey Energy Resources Program, http://energy.usgs.gov/index.html

CHAPTER 3

Sustainable Agriculture and Fishing

Agriculture has changed dramatically since the end of World War II. The food and fiber produced by American farms has contributed very significantly to the economic growth of the United States and the world. The growing, harvesting, and distribution of more food are due to new advances in technologies, mechanization, increased chemical and fertilizer use, and government policies to help farmers. Although these changes have had many positive effects, the downside includes soil erosion, overuse of fertilizers and pesticides, and groundwater depletion and contamination.

Refer to Volume III for more information about the Agricultural Revolution.

Commercial fishing has had its problems, too. Several decades of overfishing have depleted commercially important fish populations. As an example, in 2003, the Canadian government declared an end to cod fishing in nearly all of the country's Atlantic waters. Of the 15 major global ocean fishing sites, commercial fish populations in 11 areas are in decline.

Human populations are continuing to grow at the rate of about 80 million a year. Some population experts predict that the world population will reach 8 billion by the year 2025. To feed this growing population, much demand will be placed on Earth's water and land resources to grow, harvest, transport, and distribute food. With such a demand for food sources, will it be possible to develop resource sustainability in farming and fishing practices?

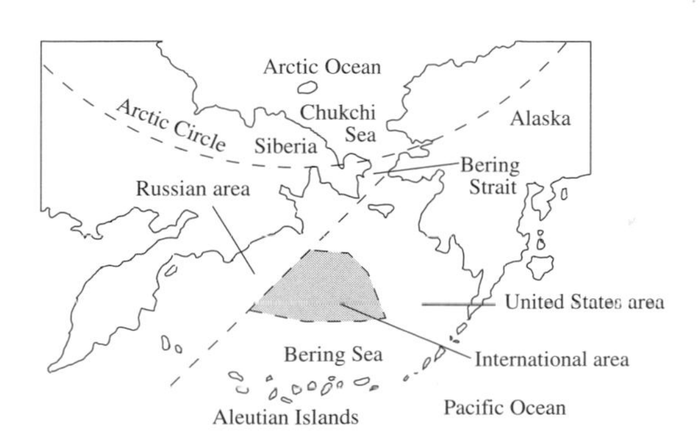

FIGURE 3-1 • Bering Sea The Bering Sea is located in the northern section of the Pacific Ocean that lies between the United States and Russia and provides 56 percent of the United States' fisheries production.

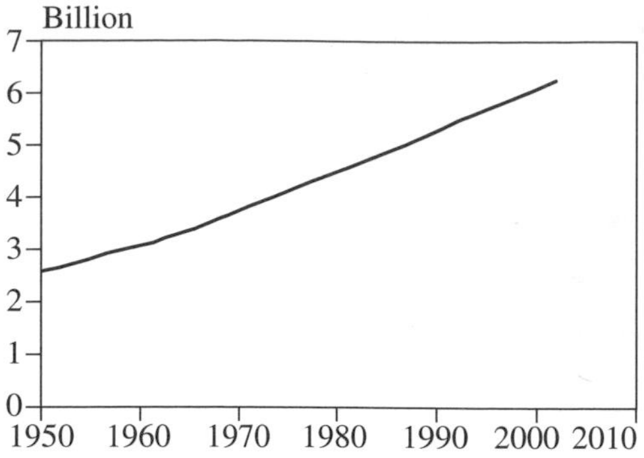

FIGURE 3-2 • World Population, 1950–2002
Source: Census Bureau

SUSTAINABLE AGRICULTURE

According to the 1990 Farm Bill, (U.S. Food, Agriculture, Conservation, and Trade Act of 1990), sustainable agriculture:

1. Integrates a system of plant and animal production practices which will, over the long term, satisfy human food and fiber needs
2. Enhances environmental quality and the natural resources, such as water and soil, upon which the agriculture economy depends
3. Makes the most efficient use of nonrenewable resources such as fossil fuels and, where appropriate, natural biological cycles and controls
4. Improves the quality of life for farmers, farmworkers, and members of rural communities, as well as society as a whole.

In essence, the Farm Bill integrates three main goals of sustainable agriculture: environmental protection, social equity, and economic profitability.

Sustainable agriculture is similar to *organic farming*, but it is conducted on a larger scale. Like organic farming, sustainable agriculture tries to establish a long-term relationship with the environment by limiting the harmful effects of more conventional farming methods. Some more practical sustainable agricultural practices include crop rotation, integrated pest management, no-till farming, crop rotation, strip cropping, and terracing.

Plant Production

CROP ROTATION

In crop rotation, different crops are alternately planted on a plot of land in order to maintain soil fertility and reduce soil erosion. Corn, tobacco, and cotton, for instance, deplete nutrients, especially nitrogen,

Crop rotation reduces the need for the extensive use of fertilizers and pesticides by alternating crops. (Courtesy of Robert Kolberg, USDA, ARS)

from the soil. On the other hand, *legumes*, such as alfalfa, barley, and beans, add nitrogen to the soil. By alternating plantings of these crops, such as corn and alfalfa, soil fertility is enhanced. Crop rotation can also suppress weeds and reduces the need for extensive pesticides and fertilizers.

No-Till Farming

In a no-till agricultural practice, the seeds of crop plants are planted along with the remains of the previous crop. Till or tillage refers to the mechanical churning up of soil, such as plowing. In no-till agriculture, soils are not tilled at the end of the growing season to remove old plants. The main benefit of no-till farming is that soils are never left bare and exposed to wind and water erosion which depletes topsoil. Even some weeds are allowed to grow. No-till farming saves time and labor because farmers do not have to plow fields. No-till farming has grown in popularity over the past decade. According to studies, in 2000, almost 20 percent of all U.S. farmland was no-till.

Other than reducing soil erosion and sedimentation, an advantage of no-till farming is that it cuts down on U.S. carbon dioxide emissions, a greenhouse gas. For example, when a field is plowed, the stirring up of the soil releases carbon dioxide into the atmosphere. The carbon dioxide in the soil is trapped by plants during *photosynthesis*. No-till farming keeps the carbon in the soil rather than releasing it into the atmosphere as carbon dioxide. Experts agree that if no-till farming practices were used 100 percent of the time on U.S. farms, carbon dioxide emissions could be cut by as much as 20 percent. One

Refer to Volume I for more information about photosynthesis.

In no-till farming, soils are not tilled at the end of the growing season to remove old plants. Plants grow among the residue of other crops. In this process, the covered soil is not left to the exposure of erosion. (Courtesy of USDA)

of the drawbacks to no-till farming is the problem of controlling the pests that live in the residue.

Contour Farming

According to some soil experts, soil forms at the rate of from 1 millimeter in 10 years to 1 millimeter in 40 years. Obviously, it takes a long time for soil to form. Sustainable farming practices, however, such as contour farming, can improve on the rebuilding of soil to 1 millimeter per year.

Contour farming is a technique in which soil is plowed according to the natural shape of the land. This practice prevents topsoil erosion from water runoff and strong winds. In contour farming, the land is plowed across, or perpendicular to, a slope to create a furrow in which the water that runs downhill will come to rest. The plowed area helps prevent the water from carrying away topsoil while providing needed water to crops. As crops begin growing, erosion from wind is also reduced because plant roots hold soil in place, and foliage serves as a protective ground cover and helps rebuild the soil.

Contour farming is sometimes carried a step further through the planting of rows of crops in strips alternating with low-growing erosion-resistant plants that help hold soil in place with their roots and foliage. This practice is generally referred to as contour strip farming. Contour farming is used as a method of soil conservation worldwide.

Strip Cropping

The agricultural technique that reduces soil erosion resulting from wind or running water is called strip cropping. In North Dakota, for example, some farms grow rows of sunflowers between wheat to help protect the wheat from wind damage.

Contour farming is a technique in which the soil is plowed according to the natural shape of the land. (Courtesy of USDA)

Strip cropping is the planting of rows of crops in strips alternating with low-growing, erosion-resistent plants that help hold soil in place. (Courtesy of Tim McCabe, USDA, NRCS)

The land on which strip cropping is practiced, is plowed in a direction that is perpendicular to its greatest wind exposure or parallel to its slope in areas vulnerable to erosion by running water. Two or more types of crops are then planted in alternating bands which run in the same direction in which the land was plowed. The crops chosen for growth and their placement are determined by when the plants are harvested. This method ensures that portions of the land remain covered by plant growth throughout much of the year. The coverage of ground by plants helps reduce erosion by slowing down high winds and water runoff, thus reducing their soil-carrying ability. In addition, the roots of growing plants help reduce erosion by holding soil in place.

Terracing is an agricultural method of reshaping the hilly terrain in order to grow crops and prevent soil erosion. (Courtesy of Lynn Betts, USDA, NRCS)

TERRACING

Terracing is a method of reshaping hilly terrain in order to grow crops and inhibit soil erosion. Terracing is commonly practiced in the mountainous regions of South America, Africa, China, Korea, Japan, the Philippines, and other places where there is heavy rainfall. In these areas, where water can flow rapidly downhill, soil erosion and contingent sedimentation in water bodies are constant problems. Terracing reduces soil erosion by slowing the water flow enough to prevent severe erosion and sedimentation.

In the early practice of terracing, the land was typically shaped into a series of broad, steplike benches. This method, called bench terracing, which is still commonly used today on steeper slopes, is very effective at reducing soil degradation because water flows gradually from one nearly level terrace to the next. Bench terracing is a good technique for growing moisture-loving crops, such as rice. When water runs down a hill, it gets trapped on each terrace. Any water that does run off simply cascades to the next terrace. In this way, large pools of water can be created for growing rice and other crops.

Most modern terracing practices involve building wide, flat terraces with shallow channels and levees which carry water at slow speeds. Such broadbase terraces are typically used on gently sloping land. Because these terraces are so wide, crops can be grown and worked with modern-day machinery.

INTEGRATED PEST MANAGEMENT

Integrated pest management (IPM) is another important part of sustainable agriculture. IPM is a set of methods used to control pest organisms including the introduction of predators, diseases, or other parasites

The farmer is applying a chemical control application to crops. (Courtesy of Jeff Varuga, USDA, NRCS)

of a pest. *Bacteria* and viruses are used to control specific pests; for example, one form of bacterial disease kills larval forms of insect pests, such as caterpillars. Bacteria are used to control Japanese beetles, and an infectious virus controls rabbits in Australia.

One practice of IPM is *biological control*, which eliminates or reduces the need for synthetic pesticides. In this method, the natural enemies of insect pests are used to control infestations on crop plants. As an example, parasitic wasps are used to control gypsy moth caterpillars. The parasites lay their eggs in the caterpillar. Over time, the wasp eggs grow and develop using the caterpillar as food. Special scents can be used to lure insects, such as Japanese beetles, into traps they find attractive.

Other techniques of IPM that help eliminate the use of synthetic pesticides include insect sterilization, which controls the growth of insect populations; planting genetically engineered crops that are resistant to insects; and using natural insecticides, such as pyrethrin, which is a liquid extract of chrysanthemum flowers.

Pest population monitoring is an essential component of IPM methodology because this practice helps to plan for the application of biological or chemical controls when pest populations are most vulnerable.

DID YOU KNOW?

Historians have reported that the Chinese used ants to control insect pests on citrus trees in A.D. 400.

Sustainable Agriculture and Animals

Using animals to aid in sustainable agriculture is a growing practice of farmers and ranchers. Instead of using chemical applications, such as pesticides and fertilizers, farmers put animals to work to control weeds and reduce insect pests and parasites. Ducks, chickens, and geese are used to eat the insects that fly above or live beneath the soil. Geese have been used to control weeds and insects in strawberry gardens. Sheep and goats are used extensively to control leafy spurge (*Euphorbia esula*), a toxic weed that is a problem to grazing cattle.

One kind of integrated pest management (IPM) is to disrupt insect pest egg development. In this photo, two entomologists are examining a papaya fruit trap that contains oriental fruit fly eggs that have been paralyzed by *Biosteres arisanus* wasps. (Courtesy of Agricultural Research Service, USDA)

Sustainable Water Uses in Farming

Sustainable farming includes management practices that improve water conservation and storage and the use of more sustainable irrigation practices, such as drip irrigation. The technology for drip (or trickle) irrigation was first developed in Israel and is, today, widely used throughout the United States, Israel, and Australia.

Drip irrigation delivers water through holes in narrow tubes directly onto the root area of individual plants at frequent intervals and in small amounts under low pressure. The slow, frequent release of water has

Drip irrigation allows a slow and frequent release of water over the soil, which helps control runoff and the rate of evaporation. (Courtesy of USDA, NRSC)

Pivot irrigation is used on a farm in southern Utah. (Courtesy of Hollis Burkhart)

several advantages over other irrigation methods. First, it reduces the total amount of water needed to irrigate crops, making this irrigation method useful in regions with limited water supplies. Second, because water is released slowly and over a small area, little water is lost to the air through evaporation, or to the soil via overspray or runoff. At the same time, fewer salts are deposited in the soil, helping to preserve soil quality. Third, drip irrigation can be used on almost all lands, regardless of their topography. The main disadvantage of drip irrigation is that the cost involved in setting up the system is higher than that of other irrigation methods; however, this disadvantage is generally offset by the benefits of drip irrigation.

IRRIGATING CROPS WITH SALTWATER

Scientists in the city of Dubai of the United Arab Emirates are testing the use of available saline (salty) water for irrigation. The United Arab Emirates, like much of the Islamic world in this region, has arid and semiarid conditions. Desertification is also a major problem in the region. As a result of these conditions, the Islamic countries are in need of freshwater alternatives for growing crops. Their research work includes irrigating vegetables, trees, and ornamental plants with saline water from the Persian Gulf. The scientists have set up a research center on a 100-hectare (250-acre) plot of land. They have collected salt-tolerant plant species from arid lands and have stored more than 6,000 plant species as seed materials which will be planted and tested in the future. Currently, they are experimenting with a variety of salt-tolerant plants and irrigation systems on a 35-hectare (85-acre) farm. More research needs be conducted, but the scientists at the center are optimistic about using saline water for irrigation in the future.

Some agricultural economists agree that current agricultural strategies are too heavily dependent on nonrenewable fossil fuel energy, especially petroleum. In a sustainable agricultural system, there must be a reduction in nonrenewable energy sources and a substitution of renewable sources, such as wind or solar energy.

BIOTECHNOLOGY, BENEFITS, AND CONCERNS

Biotechnology is the use of technology to manipulate different genetic materials or biological features to produce living organisms or biological agents with economically desirable characteristics. Biotechnology

Agroecology

The science that applies principles of ecology to agriculture is called agroecology. The goal of agroecology is to develop sustainable, or long-term, agriculture practices that will produce quality agricultural products without harming soil, organisms, or other aspects of the environment. Such practices make use of biological controls (living organisms) such as IPM in place of chemical pesticides which can decimate organisms other than their targets. Other goals of agroecology include making use of nature's biogeochemical cycles and practicing energy and soil conservation techniques. Agroecology also strives to improve irrigation practices to help conserve water resources and to reduce soil damage, such as salinization, which results from irrigation. Using such knowledge, farmers and ranchers can reduce their use of agrochemicals and keep soil healthy while reducing agricultural pollution and conserving natural resources. Techniques of agroecology have been implemented with positive results in many parts of the world. For example, Chinese cotton farmers have reduced pesticide use by as much as 50 percent using biological controls. Many rice farmers in Indonesia turned to IPM after most of the pesticides used on this crop were banned by the government. In both countries, crop yields increased or remained the same, but the costs involved in growing the crops decreased.

examples would include artificial insemination of cattle and the developing of pesticides based on natural toxins. Biotechnology can be used to develop raw materials for renewable energy sources such as ethanol.

In agriculture, biotechnology offers a great potential to lower food costs, enhance food quality, and provide safer foods. Food industry experts estimate that about 70 percent of all foods sold in U.S. stores contain genetically modified ingredients. In fact, the United States leads the world in the production of genetically engineered foods.

Many world leaders and scientists believe that the production of biotech foods is one way to get nutritious food to starving people. Some 800 million people go hungry each day according to the United Nations Food and Agriculture Organization (FAO). Biotechnology would allow farmers to grow more food on existing arable land, even in poorly developed land areas, saving the need to put more wilderness areas into farming activities.

Although plant technology is an important factor in reducing hunger, there are concerns about environmental safety. Trying to contain and control genetic modifications introduced into the environment may be difficult if not impossible. Critics argue that we should be exploring solutions to world hunger other than biotech foods.

Percy Julian (1899–1975) a pioneer in biotechnology spent much of his career studying soybeans. He discovered several chemicals from soybeans that are used to treat health problems such as certain cancers and eye disorders. More than 300 food products are made from soybeans. (Courtesy of Agricultural Research Service, USDA)

SUSTAINABLE FISHING AND AQUACULTURE

DID YOU KNOW?

Until the collapse of the Canadian cod fishing industry, Canadian fishers had been harvesting cod for more than 500 years in the waters off Newfoundland in the Atlantic Ocean.

Records kept by the United Nations FAO have reported that of the world's 15 main fishing regions, 4 are depleted and 11 are declining. Several decades of overfishing in most of the world's major fisheries have pushed many commercially important fish populations into steep declines. According to the National Marine Fisheries Service, 80 percent of the known commercial valuable fish populations are overexploited. These include bluefin tuna, swordfish, oysters, and red snapper. The collapse of the Canadian cod fishery in the early 1990s left 30,000 commercial fishers unemployed. Some commercially important stocks are in such a critical state that all fishing has been shut down, or sharply curtailed.

The Issues

To keep up with population growth, commercial fisheries are becoming more reliant on increasingly sophisticated fishing gear, such as large nets and harvesting methods. There are more fishing vessels than ever,

TABLE 3-1 Commercial Fishing

Region	Gross Tonnage (1,000 metric tons) 1970	1992	Percent Growth, 1970–92
Asia	4,802.3	11,012.5	129
Former Soviet Union	3,996.7	7,765.5	94
Europe	3,097.4	3,018.3	−3
North America	1,076.9	2,560.0	138
South America	361.5	816.5	126
Africa	244.0	699.1	187
Oceania	37.1	122.3	230
World	**13,615.9**	**25,994.2**	**91**

The gross tonnage of fish caught between 1970 and 1992.
Source: Food and Agriculture Organization of the United Nations.

Commerical fishing is an important part of the world's economy. (Courtesy of USDA)

including giant "factory ships," which search the seas for fish with sonar and spotter aircraft. These vessels, equipped with sophisticated, onboard electronic tools and satellite data receivers, can track large schools of fish day and night.

Harvesting fish has also improved with the use of long drift nets, longliners, and purse seines which can catch thousands of kilograms (pounds) of fish in a brief period of time. These oversized nets allow a small fishing crew to haul in tons of fish. The nets also catch many other species, known as bycatch, which are not wanted by the fishers and are discarded.

Refer to Volume IV for more information about overfishing and loss of ocean species.

Creating More Sustainability Programs

The National Marine Fisheries Service and other government agencies, state fisheries, commercial fishers, environmental groups, and others are working on a variety of issues to bring about more sustainable fishing practices. One strategy for the fishing industry to increase fish and shellfish production is to establish fish farms, called aquaculture. Aquaculture is a commercial enterprise including the feeding and raising of fish and shellfish in freshwater and saltwater ponds, enclosed lagoons, and open, offshore waters. The United Nations FAO estimates that at some time between 2015 and 2025, half of all fish consumed in the world will come from aquaculture enterprises.

Aquaculture

Aquaculture is the cultivation of fish, shellfish, or aquatic plants in natural or controlled marine or freshwater environments. Today, aquaculture is a multimillion dollar business. Much of the trout, catfish, and shellfish consumed in the United States are products of aquaculture. One report estimates that aquaculture produces more than 25 million tons of fish and shellfish a year. Other estimates report that about 20 percent of all commercial fish are raised in an aquaculture environment.

Cracking Down on the Fin Trade

Certain shark species are hunted specifically for their fins, which are used in making sharkfin soup, a popular delicacy in Asian countries. According to a marine conservation group, fishing fleets take an estimated 100 million sharks annually. However, much of the fish, other than the fins is thrown away. Although the flesh can be sold, it nets very little money for the fishers. The fins, however, can be sold for as much as $400 per kilogram (2.2 pounds). After the fins have been removed, the carcass is pitched overboard. Conservationists are complaining about the unsustainable harvesting of sharks and warn that some shark populations may collapse. They propose quotas for individual species. U.S. vessels are enforcing the U.S. Shark Finning Prohibition Act, which bans foreign vessels in U.S. territorial waters from possessing fins unless the rest of the shark's carcass is also onboard.

There are several reasons behind the tremendous growth and interest in aquaculture. The main one, perhaps, is the recognition that the world's oceans, lakes, and rivers cannot produce enough food to satisfy the world's appetite for fish and other types of seafood.

Aquaculture encompasses a wide variety of activities, including cultivation of fish, such as catfish and trout, for food; the rearing of ornamental fish, such as carp and koi, for aquariums; the raising of bait fish for the fishing industry and sporting fish for restocking lakes and ponds; the cultivation of oysters for obtaining pearls; the cultivation of mussels for food; and the growing of seaweed for food.

Today, fish farming, an important industry in the United States, Philippines, Japan, China, India, Israel, and Europe, is by far the most common form of aquaculture. Fish farming is the practice of raising fish in captivity in order to improve their growth and reproduction, similar to the way livestock is raised on land. Most fish farms consist of many enclosures, ponds, lakes, tanks, pens, and long, narrow channels, each containing fish at varying stages of development. Fish farmers manage the aquatic environments by circulating clean water through the pens and tanks and by protecting the fish from predators, disease, and parasites. In these types of farm operations, fish are grown to maturity and then harvested for food.

Another type of fish farm common in the United States is known as a fish ranch. Here, many fish, particularly sport fish species, such as salmon, are hatched in small ponds and then released into rivers. The fish then migrate downstream to the ocean where they will reach adulthood. Once these fish mature, they instinctively return to the river from which they were released in order to reproduce. When they do so, they are captured and harvested for food.

Plant Aquaculture

The aquatic plants raised in aquaculture include ornamental plants, such as pond lilies, and native species of plants used for habitat restoration. The vast majority of the plants systematically grown in aquaculture operations are seaweed, a type of *algae*. The cultivation of seaweed is particularly popular in China and Japan, where it is an important food source.

Since the seventeenth century, for example, Japanese aquaculturalists have grown their own seaweeds. Traditionally, farmers cultivated the algae by placing long bamboo sticks into rivers. When small seaweed plants began to grow, the sticks were removed and brought to the sea, where the plants thrived in a mixture of freshwater and saltwater. Today, the cultivation of algae in Japan and elsewhere is highly mechanized.

Another force behind the growth in aquaculture is the increased interest, particularly within the United States, in eating a healthful diet. Numerous studies have concluded that fish and seafood are low

The vast majority of plants grown in aquacultural farms are seaweed, a type of algae, which is an important food source in Asia. (Courtesy of Agricultural Research Service, USDA)

in sodium, fat, and *cholesterol*. Additional studies have found that certain fish contain fatty substances (oils/fatty acids) that have the effect of reducing cholesterol in the body.

Environmental Concern of Aquaculture

There are, nevertheless, problems with some of the fish farms, particularly shrimp-raising farms. Although aquaculture is a possible long-term solution to global fishing problems, the practice is not always conducted in a sustainable way. Disease outbreaks, chemical pollution, and the environmental destruction of marshes and mangroves have been linked with fish farm activities. Unsustainable shrimp and salmon aquaculture, in particular, can deplete native fisheries, disrupt coastal ecosystems, and pollute the ocean with excess nutrients and pesticides. Environmentalists believe that more sustainable aquacultural practices are needed to control the potential for pollution and damage of natural resources.

Refer to Volume IV for more information about the environmental concerns of aquaculture.

Rebuilding Fish Stocks

The federal government and nonprofit environmental organizations, such as the Environmental Defense organization, are working on management systems to rebuild fisheries. One plan would set a limit on the number of fishers eligible to work in threatened areas. In this plan, the fishers would have a trading program that would enable them to buy and sell the limited fishing rights among themselves.

The Environmental Defense organization also has a plan called individual fishing quotients (IFOs). In the plan, fishers are assigned individual shares of the annual allowable catch. The plan has been successfully used in more than 100 fisheries worldwide. Fishers decide when weather and market conditions are favorable for harvesting fish. The plan eliminates incentives or rewards to overfish. As the fish stocks increase, the allowable catch also increases. Since all of the fishers have shares, they can sell their shares to another if they want to leave the business. The Environmental Defense plan can be used for threatened species such as Pacific groundfish, Alaskan crab, Gulf shrimp, and New England scallops.

The National Marine Fisheries Service advocates a plan similar to the IFOs. Their system is called the individual transferable quotas (ITQs). In this plan, permits are issued to fishers to allow them to harvest a certain total catch of fish per year. The fishers can lease or sell their permits to other fishers. This program has been successful in several U.S. fisheries.

Eco-labeling of Fish

Another practice to encourage sustainable fishing practices is to label any fish and fish products that have not been overfished or harvested in ways that harm the ocean ecosystem. One of the first groups to certify sustainable fish products was the Marine Stewardship Council (MSC) based in London, England. Consumers in the United Kingdom are now able to buy fish products bearing the MSC label.

The MSC announced that Alaskan salmon is the first U.S. fish stock to be certified as sustainable. Alaskan salmon are eligible to bear the MSC label, informing consumers that, when they buy such MSC-labeled seafood, they are supporting healthier oceans and a healthier environment.

The MSC's globally respected certification program provides its label to fisheries that meet strict, peer-reviewed standards of sustainability. The eco-label allows consumers quickly to identify the best environmental choices in seafood—fish that have not been overfished.

In the coming years, fish marked with the MSC label will be available in seafood restaurants and supermarkets. By buying only fish and seafood with the MSC eco-label, everyone can directly contribute to the conservation of our oceans.

FIGURE 3-3 • The Marine Stewardship Council label is used to label fish and fish products that have not been overfished or harvested in a way that harms the ocean ecosystem. (© Marine Stewardship Council, license code MSCI0199)

Other Measures

Other measures may be effective in curbing overfishing and bringing fish stocks back to acceptable levels. In one FAO study, "The State of World Fisheries and Aquaculture (SOFIA)," it was suggested that even though fish stocks are currently below their sustainable productivity levels, it might be possible to return to these levels by reducing fishing efforts. This can be achieved by increasing the age of fish at first capture, prohibiting the exploitation of juvenile fish, increasing mesh sizes, and temporarily or permanently closing those areas where young fish are concentrated.

Other plans for attaining a more sustainable fishing management program include

- Cutting down on the number of fishers and fishing fleets
- Improving the science of collecting and assessing the data of threatened species
- Reducing the bycatch and nontargeted fish in the harvesting of targeted species
- Developing more cooperation between commercial fisheries, federal and state agencies, land management organizations, and environmentalists.

According to the National Oceanic and Atmospheric Administration, there have been some successes in restoring fish stocks:

- The collapsed striped bass fishery off the Atlantic coast has recovered, and widespread fishing is once again allowed.
- The Atlantic group of Spanish mackerel, heavily overfished, is recovering, permitting larger commercial and recreational harvests.
- The agency's management of North Pacific groundfish has kept the fishery the most productive and wealthiest in U.S. waters.
- The tuna and swordfish stocks fished in Western Pacific waters under U.S. jurisdiction remain healthy.
- The agency has made significant progress on restoring many depleted fish stocks, such as New England groundfish, Gulf of Mexico red snapper, and Atlantic bluefin tuna.

Vocabulary

Algae Group of one-celled free-floating green plants.

Bacteria Microscopic organisms found in water, soil, air, and in living and nonliving organic matter. Some cause disease.

Biological control Elimination or reduction of pests, usually insects, by using predators, parasites, or pathogens.

Cholesterol Substance found in fats and oils and produced by the liver; it is an essential part of all cells. High levels of cholesterol can be deposited in the walls of the arteries causing blockage.

Legumes Type of plant that produces seeds in pods. Some are valuable because they have root nodules containing nitrogen-fixing bacteria.

Organic farming Growing food naturally without the use of synthetic pesticides or fertilizers.

Photosynthesis The process by which sugar is manufactured in plant cells.

Activities for Students

1. Visit a local farmer's market and interview an organic farmer. How does an organic farmer grow his or her food differently from a non-organic farmer? What made the farmer interested in organic agriculture?
2. There has been much discussion around the world about the uses of biologically engineered crops. Hold a debate at your school addressing the benefits and disadvantages of using science to change the makeup of plant seeds genetically.
3. Think of an eco-label for one of the foods you like, such as cereal. Then research food companies who produce the product in a sustainable manner.

Books and Other Reading Materials

Cone, Molly. *Come Back, Salmon: How a Group of Dedicated Kids Adopted Pigeon Creek and Brought It Back to Life*. San Francisco: Sierra Club Juveniles, 1994.

Helvarg, David. *Blue Frontier: Saving America's Living Seas*. New York: W. H. Freeman, 2001.

Paladino, Catherine. *One Good Apple. Growing Our Food for the Sake of the Earth*. Boston: Houghton Mifflin, 1999.

Winckler, Suzanne. *Our Endangered Planet: Soil*. Minneapolis: Lerner Publications. 1993.

Websites

Agricultural Research Service (ARS), http://www.ars.usda.gov/

Agriculture Network Information Center, (AgNIC), http://www.agnic.org/

Alternative Farming Systems Information Center, http://www.nal.usda.gov/afsic/

American Oceans Campaign, http://www.americanoceans.org

Center for Marine Conservation, http://www.cmc-ocean.org

Earth Island Institute, http://www.earthisland.org

Marine Stewardship Council, http://www.msc.org

National Marine Fisheries Service, http://www.nmfs.gov

National Oceanographic and Atmospheric Administration Fisheries, http://www.nmfs.gov/

SeaWeb, http://www.seaweb.org

United Nations Food and Agriculture Organization Fisheries, http://www.fao.org/waicent/faoinfo/fishery/fishery.htm

USDA Sustainable Development and Small Farms, http://www.usda.gov/oce/sdsf/

U.S. National Agricultural Library, http://www.nal.usda.gov/

U.S. National Cooperative Soil Survey, http://www.statlab.iastate.edu/soils/nsdaf/

CHAPTER 4

Sustainable Forests and Preserving Wildlife Species

The world's forests cover almost one-third of Earth's land surface, and they are important natural resources for people. Each year, millions of hectares (acres) of forest are cut down for timber production, for fuelwood, and for land development, including housing, road building, farming, and other human activities. Forestry is an important part of the economy; the products of the forest provide us with food, shelter, employment, and energy. Excessive logging and inadequate forest management practices are causing environmental problems. The degradation of the environment, including the loss of forests, devastates wildlife species and their habitats. Developing short-term and long-term strategies to protect forests and preserve wildlife species will be essential to the sustainability of these natural resources.

DID YOU KNOW?

Recent surveys show that the prospect of forest loss has overtaken recycling and chemical use as a key environmental issue for the general public.

Clearcutting is a harvesting method of cutting down all trees and vegetation in an area. (Courtesy of USDA)

DISAPPEARING FORESTS

Forest loss has occurred catastrophically throughout geologic records. Natural climate change has accounted for most forest losses; however human activities, including logging, forest fires, and land-clearing activities, have reduced large tracts of forest areas, particularly in the rain forest. The pace of forest destruction has accelerated in the 1990s and it continues to rise; currently more than 400,000 hectares (950,000 acres) of forest are cleared or degraded every week. Researchers believe that more than 8,750 of the 80,000 to 100,000 species of trees known to science are disappearing.

The major threat to the world's forest is the permanent removal of trees and vegetation, or *deforestation*. Deforestation is most severe in the tropical rain forests of Africa, Asia, and Central America. The United Nations Food and Agricultural Organization (FAO) estimates that Central America has lost more than 65 percent of its forests; central Africa, more than 50 percent; and Southeast Asia and South America, more than 30 percent. Much of this deforestation is the result of harvesting forests to provide timber and wood products for lumber, paper, and fuelwood. Deforestation is also a result of clearing forests for agricultural and ranching activities, particularly in Nepal and Brazil.

DID YOU KNOW?

The United States is the fourth most forested country, exceeded by the Russian Federation, Brazil, and Canada.

Refer to Volume IV for more information about deforestation.

Benefits of Forests

Trees in forests are critical to the welfare of our planet and play a vital part in controlling climate and water cycles. They help keep the air clean by filtering pollutants and reduce the risk of global warming by absorbing carbon dioxide and other greenhouse gases. Forests also act as *watersheds*. Forests absorb rainfall and slowly release it into streams and rivers, moderating both floods and droughts, and regulating water

FIGURE 4-1 • Rhine River Watershed The Rhine River watershed covers an area of 190,000 square kilometers and includes a wide variety of riparian habitats and ecosystems for many plant and animal species. The river is also an important drinking water supply for eight million Europeans. Several countries bordering the river have set up a number of programs to clean up the river.

flows. Forests keep soil erosion in check, which prevents *siltation* of waterways and damage to coral reefs, fisheries, and spawning grounds.

People who live in and near forests depend on them for much of their food, medicine, clothing, and timber; most of these products come from the plants and trees that grow best in natural forests. For example, the rain forest is the most abundant source of medicinal plants used today in making medicines. One-fourth of the medicines available are derived from plants in the rain forests. Chemicals from

FIGURE 4-2 • Years ago, quinine, an extract from the bark of the Cinchona tree, was used as an aid in the cure of malaria. Now quinine is made synthetically. Today, however, the quinine bark is used in a prescription drug that treats irregular heart rhythms and as a flavor ingredient in soft drinks. Quinine comes from the bark of the Cinchona which comprises about forty species of trees that grow up to 15 to 20 meters (45 to 60 feet).

FIGURE 4-3 • Suma is a large, shrubby ground vine found in the tropical rainforests of Brazil, Venezuela, and Peru. It is used in the treatment of asthma, leukemia, and high blood pressure.

rain forest plants are used in surgery and for internal medicine. These medicines are used to treat illnesses ranging from headaches to contagious diseases such as malaria. Other products of the forests include industrial commodities such as rubber, waxes, and fibers.

Refer to Volume II for more information about forests and forests products.

MANAGING FORESTS SUSTAINABLY

Sustainable forest management involves the managing of the forest in such a way that it meets the economic, environmental, and social requirements of the community. In general, sustainable forest management is the ability to grow successful crops indefinitely and harvest forest products (fuel, fruits, lumber, paper, and so on) without causing any adverse efforts to the biodiversity of the forest.

The management of a forest, including the growing and cultivation of trees as a crop, is called silviculture. Silviculture includes planning, planting seeds or seedlings, fertilizing the land, and harvesting the entire trees. Without appropriate silviculture, sustainable forest management is impossible.

Tree Harvesting Methods

Present-day forest managers use a variety of tree harvesting methods to minimize habitat destruction and soil erosion while protecting the ecosystems. Some of the methods include shelter-wood harvesting, seed-tree cutting, selective cutting, and strip cutting. These four techniques are more sustainable and have less impact on a forest than *clear-cutting* practices.

SHELTER-WOOD HARVESTING

Shelter-wood harvesting is a sustainable development practice used by the timber and logging industries. In this technique, all mature and fully grown trees are removed from the forest. Generally, trees are removed over several decades. As the taller, more mature trees, called shelter wood, are removed, the shade once created by their canopies is also eliminated. Thus, younger, shorter trees receive increased exposure to sunlight, helping to promote their growth. Shelter-wood harvesting cuts down on erosion and reduces habitat destruction.

SEED-TREE CUTTING

Seed-tree cutting is a type of selective harvesting or selective cutting in which mature, economically desirable trees are left standing in an area that has been cleared of other trees. The mature trees, called seed trees, are left uncut as a source of seeds to replenish a new crop of trees. Seed-tree cutting is a sustainable forestry practice designed to replenish tree populations in forest areas where trees are harvested as a source of timber.

Selective Cutting

In selective cutting, only the mature trees are marked and cut down; the immature ones are left to grow. Selective cutting helps preserve the diverse species in the forest and protects them from alien species as well. In this method, the remaining trees of different sizes and types are left to maintain the ecosystem. Selective cutting is used in mixed, diverse forests such as those in the northeastern United States. A mixed, diverse northern hardwood forest would include maples, oaks, hickories, elms, and some cone-bearing pines. Foresters cut down and remove desirable trees, such as maples, as well as sickly and unhealthy trees. There is some criticism to this method because some of the trees that are not harvested are damaged during the cuttings.

Strip Cutting

In strip cutting, another harvesting method, trees are cut in narrow rows or strips. The uncut areas remain open for use in recreation and for wildlife habitats. The trees that are not stripped provide seeds and protect young trees from sun and wind. Strip cutting leaves the forest intact for natural reseeding.

Reforestation

Once the trees are harvested, it is important to plant new trees to reforest the land area. Reforestation is now used in many countries to replace forests that were clear-cut mostly to produce land for agriculture. In recent years, Israel has conducted reforestation projects to restore the forestland in that country and to slow down the process of deforestation and desertification. The projects have led to almost 5 percent of

In selective cutting, only the mature or older trees are cut down while the young trees are left standing to grow to maturity. (Courtesy of Bruce Hanson, USDA, NRSC)

that nation's land being occupied by natural woodlands and reforested areas. In South Korea, communities have been reforesting at the rate of 40,000 hectares (98,000 acres) a year. In Kenya, environmentalist Wangari Mutta Maathai began the Greenbelt Movement in the 1970s as a means of restoring the forests in that country. Maathai's efforts were designed to create jobs, mostly for women. Since it first started, the Greenbelt Movement has been quite successful, and it has now expanded to several other countries on the African continent.

Tree Farms

Tree farms can also be part of a reforestation program, particularly in areas that have been devastated by clear-cutting techniques. A tree farm or tree plantation is an area of land on which trees are grown and managed for commercial uses, such as fuelwood, pulpwood for paper uses, Christmas trees, and ornamental decorations. Currently, most tree farms are found in temperate climates; however, more and more tree farms are being established in tropical regions where trees can grow much faster than in cooler climates. Tree farms can also provide habitats for a variety of wildlife species.

DID YOU KNOW?

One of the earliest examples of tree farming occurred in the late 1800s when British scientist Henry Wickham collected more than 70,000 seeds from the rubber plant Hevea brasiliensis in the Amazon jungle for the purpose of growing trees in Kew Gardens, London, England. Later, the saplings were transported to Ceylon, India, and Malaya where they were used to establish the rubber industry.

Certifying Sustainable Forest Products

Many private forestry corporations can be certified as sustainable if they apply sustainable forestry practices in harvesting their trees. The certification informs the consumer that the wood or wood product sold was produced from well-managed forests without damage to the environment. One independent certifier is the Forest Stewardship Council (FSC).

As a result of the certification programs, some retailers, such as Home Depot, have eliminated all noncertified wood products in their stores. Home Depot is the world's third largest lumber retailer selling certified wood products. To date, about 179 companies throughout the United States carry FSC chain-of-custody certification, and 52 U.S. forest management companies are FSC certified.

Refer to Chapter 6 for more information about sustainable consumer products.

Several forestry associations have taken steps to provide sustainable forest practices. The American Forest and Paper Association (AFPA), which represents 95 percent of the industrial forestland in the United States, has established the Sustainable Forestry Initiative Principles and Guidelines (SFI). The guidelines include performance measures for reforestation and the protection of water quality, wildlife, visual quality, biological diversity, and areas of special significance. The National Woodland Owners Association, together with the Association of Consulting Foresters, has accepted sustainable forest management as a goal. This is reflected in their Green Tag Program, which certifies

DID YOU KNOW?

Trees help reduce global warming by removing carbon dioxide from the atmosphere. Carbon dioxide is a major contributor to global warming.

Switching to Nonwood Alternative Papers

Paper is the fastest growing segment of the wood product industry. One estimate indicates that one out of every three harvested trees is used as pulp to make paper. One way to cut down on the use of trees for paper pulp is to develop more tree-free paper fibers.

Currently, there are several tree-free plant fibers, including bagasse, kenaf, flax, bamboo, and hemp. Some farmers in Georgia, Texas, and Mississippi are growing kenaf, a cane-like plant that is a native of Africa. Kenaf is used to make paper for the printing industry, for animal bedding, and as an absorbent paper to clean up oil spills.

According to one papermaker, the manufacturing of kenaf paper uses less energy and fewer chemicals to prepare the pulp of the plant for papermaking than conventional processes. Bagasse, one of the world's most widely used nonwood fibers, is the crushed outer stalk left over after the juice is squeezed out from sugar cane. Bagasse is used to make paper towels and tissues. The Kimberly Clark papermill is one of the major producers of bagasse paper.

Hemp, bamboo, and even parts of the banana plant have been converted to paper pulp to produce paper for textbooks, notebooks, and other paper products. Flax is used to make currency bills, airmail packages, and even teabags. High-quality writing paper is also made from flax. Consumers now have the option to use tree-free papers wherever possible to help cut down on the use of wood-pulp papers.

FIGURE 4-4 • More than 1,000 middle schools in Japan grow, study, and make paper from the kenaf plant each year. Kenaf is a cane-like plant that is an alternative tree-free plant fiber. Kenaf is used to make a variety of paper products.

wood products produced by small, nonindustrial wood producers using sustainable practices.

Refer to Volume II for more information about papermaking.

PRESERVING WILDLIFE SPECIES AND PROTECTING BIODIVERSITY

Disappearing Wildlife Species

Many environmentalists confirm that by the year 2100 between 20 and 50 percent of all wildlife species which lived on Earth since 1900 will become extinct. One environmental study reported that as many as 100 species disappear from Earth each day. Some scientists have estimated that wildlife species are becoming extinct faster today than at any other time in Earth's history.

Wildlife species can become extinct as a result of climate change and other natural disasters such as droughts, landslides, earthquakes,

DID YOU KNOW?

During the Permian geological period, about 250 million years ago, between 50 and 90 percent of all species were terminated. This was a time of mass extinctions.

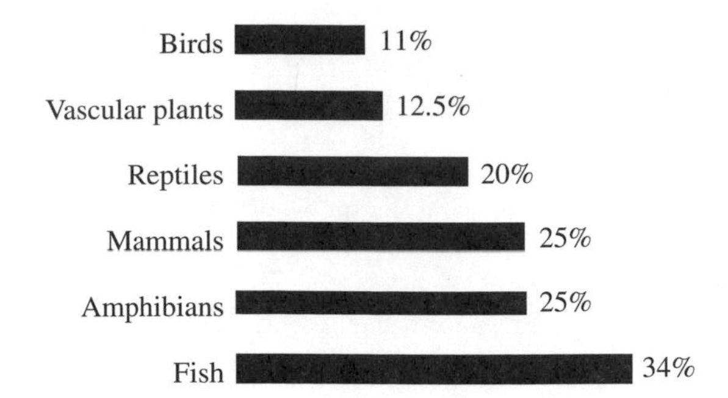

FIGURE 4-5 • Wildlife under Pressure: Percentage of Species Threatened, 1996

volcanic eruptions, and meteor impacts. These events can destroy species and their habitats. Human activities also have an impact on wildlife species. Some of these human activities include

- Releasing pollutants into air, water, and soil which damage and destroy habitats
- Diverting water from rivers and streams to other areas to provide drinking water and irrigation
- Deforestation, particularly in rain forests
- Clearing land to develop roads, farmland, grazing pastures, and housing developments.

Refer to Volume IV for more information on habitat loss.

Habitat loss accounts for 85 percent of all species that are threatened with extinction. Overhunting and poaching are also major problems in some countries.

Species which are at risk of extinction include endangered species and threatened species. The Fish and Wildlife Service defines endangered species as those at immediate risk of extinction which probably will not survive without direct human intervention. Threatened species are those which are abundant in parts of their range, but are declining in total numbers and thus are at risk of extinction in the foreseeable future.

Protecting Biodiversity and Wildlife Species

BIODIVERSITY TREATY

Several treaties are in effect to protect biodiversity and wildlife species. The Biodiversity Treaty is an international agreement developed to protect Earth's biodiversity. The treaty resulted from the United Nations Earth Summit held in Rio de Janeiro, Brazil, in 1992, where it was signed by more than 150 nations. The major provision in the Biodiversity Treaty is for the establishment of a worldwide inventory of threatened and endangered species and for cooperation among nations to protect such species. The treaty also provides financial assistance from wealthy, developed nations to less wealthy, developing nations to help them protect potentially valuable species.

Refer to Volume I for more information about biodiversity and wildlife species.

CITES

Another international environmental treaty that protects wildlife species is the Convention on International Trade in Endangered Species of Wild Fauna and Flora (CITES). This 1973 international treaty protects more than 600 species of animals and plants, including the gorilla and rhinoceros. The early 1990s saw some success achieved in prohibiting trade in rhinoceros horn, elephant ivory, and endangered orchids. In many countries, however, a lack of local law enforcement, the willingness of some individuals to trade in endangered species, and the activities of poachers and traders put the future of many species in jeopardy despite the existence of legal protections.

Wilderness Act of 1964

The United States has also been active in protecting wildlife species and biodiversity. The Wilderness Act of 1964 was passed by the U.S. Congress to protect and preserve more than 38.5 million hectares (95 million acres) of federal land. The land, designated as "wilderness areas," is used entirely to sustain and protect biodiversity and for educational and research purposes. Under this law, the land shall remain as a preserve—unimpaired for future use. The law prohibits permanent structures, roads, motor vehicles, timber harvesting, and mining on these lands. Some recreational activities, such as hiking and similar activities, are allowed.

Endangered Species Act

In 1973 the U.S. Congress passed the Endangered Species Act (ESA). This act requires the conservation of threatened and endangered species and the ecosystems upon which they depend. The act also discourages the *exploitation* of endangered species in other countries by banning the importation or trade of any endangered species or any product made from such species.

The U.S. Fish and Wildlife Service (FWS) and the National Marine Fisheries Service are the two federal agencies in charge of managing the ESA. Their responsibilities include working with private landowners, citizens, and organizations to

- Conserve species
- Determine which species need protection
- Restore listed species to a secure existence or recovery.

The two federal agencies cooperate with private landowners. Because approximately 70 percent of all endangered and threatened species inhabit privately owned lands, the cooperation and involvement of the landowners are crucial to the management of at-risk species.

ENDANGERED SPECIES AND THREATENED SPECIES LIST

A major function of the FWS is to identify and recover endangered species. The FWS leads the federal effort to protect and restore the animals and plants that are in danger of extinction, both in the United States and worldwide. Using the best scientific information available, the agency identifies the species that are or have the potential to become endangered or threatened. The species that meet the criteria of the Endangered Species Act are placed on the Interior Department's official List of Endangered and Threatened Wildlife and Plants.

As of December 31, 2001, the FWS included 386 animal species and 595 plant species of the United States on its list of endangered species. The number of animal species from each group include 64 mammals, 78 birds, 14 reptiles, 11 amphibians, 71 fishes, 62 clams, 21 snails, 35 insects, 12 arachnids, and 18 crustaceans. The number of U.S. plant species listed as endangered includes more than 500 species of flowering plants, a few conifers, and about 30 species of ferns or other plants. In addition, 128 animal species and 145 plant species are identified as threatened species.

A more complete survey of plant and animal species worldwide which are recognized as endangered or threatened is maintained by the World Conservation Union (IUCN), the world's largest species survival commission. The IUCN publishes a Red List of Threatened and Endangered Species. According to their Red List statistics, worldwide, 11,096 species (5,485 animals and 5,611 plants) are critically endangered, endangered, or vulnerable. Of these, 1,939 are critically endangered (925 animals and 1,014 plants).

TABLE 4-1 Some Animals Appearing on the First Endangered Species List (1967)

Mammals	**Birds**
Indiana bat	Aleutian Canada goose
Timber wolf	Hawaiian duck
Red wolf	California condor
Grizzly bear	Florida Everglade kite
Black-footed ferret	Southern bald eagle
Florida panther	**Reptiles**
Florida manatee	American alligator
Key deer	Blunt-nosed leopard lizard
Fishes	San Francisco garter snake
Shortnose sturgeon	Texas blind salamander
Montana westslope cutthroat trout	
Gila trout	
Humpback chub	

THE RED LIST OF ENDANGERED SPECIES

The Red List is similar to, but broader in scope than, the U.S. Fish and Wildlife Service Endangered Species. The primary purpose of the Red List is to increase global awareness of species that are at risk of going extinct so actions may be taken to prevent such occurrences.

Red Lists are published every few years with separate lists being prepared for plants and animals. The lists focus on identifying species

FIGURE 4-6 • The Florida panther lives in the subtropical forests of Everglades National Park in southern Florida. It was on the first list of U.S. Endangered Species in 1967.

TABLE 4-2 Some Endangered and Threatened Species

Group	Endangered		Threatened		Total
	U.S.	Foreign	U.S.	Foreign	species
Mammals	61	251	8	16	336
Birds	75	178	15	6	274
Reptiles	14	65	21	14	114
Amphibians	9	8	8	1	26
Fishes	69	11	41	0	121
Clams	61	2	8	0	71
Snails	18	1	10	0	29
Insects	28	4	9	0	41
Arachnids	5	0	0	0	5
Crustaceans	17	0	3	0	20
Animal subtotal	357	520	123	37	1,037
Flowering plants	540	1	132	0	673
Conifers	2	0	1	2	5
Ferns and others	26	0	2	0	28
Plant subtotal	568	1	135	2	706
Grand total	925	521	258	39	1,743

Source: United States Fish and Wildlife Service, 1998.

threatened with global extinction. Beginning in 1994, the IUCN created several classification categories for the organisms on its lists. The categories are extinct, extinct in the wild, critically endangered, endangered, and vulnerable.

- Extinct species are those in which there is no doubt that all individuals of that species have ceased to exist.
- Extinct in the wild is used to classify organisms that are believed to exist now only in captivity (zoos, wildlife refuges, captive breeding facilities) when cultivated (grown in nurseries or on croplands), or as a naturalized population. A naturalized population is one that survives outside its past range.
- Critically endangered is used for classifying organisms that are at high risk of extinction in the immediate future.
- Endangered are those species populations that are facing a high risk of extinction in the wild, but with no timetable for extinction.
- A species is deemed vulnerable when it is not critically endangered, but poses a risk of extinction in the near future. The vulnerable species category is similar to the threatened species designation used by the U.S. Fish and Wildlife Service.

The Red List also documents how human activities, such as the introduction of alien species into an ecosystem, pollution, and the fragmentation or loss of habitats, have affected various wildlife populations. The main areas where mammals, birds, and plants (trees) seem to require the most conservation effort, according to the IUCN, are in the tropics, including Brazil, Colombia, Ecuador, Mexico, East Africa (Tanzania), and the Southeast Asian countries of China, India, Indonesia, and Malaysia. For mammals alone, the situation appears most critical in Africa; for birds, Argentina and the Southeast Asian block of Myanmar, Vietnam, and Cambodia are the critical spots.

PROGRAMS TO PROTECT WILDLIFE SPECIES AND HABITATS

To maintain Earth's biodiversity, scientists recommend that people learn as much as they can about the environment and use its resources wisely. This may help prevent the extinction of many species in the future. In addition, several programs have been established to protect those species that have been identified as being on the verge of extinction. Such programs can be expensive to implement and do not always meet with success. However, until all of Earth's organisms are known and their roles in the environment understood, not making such efforts may prove to be even more costly to the future of humans and the environment. Wildlife refuges and *captive propagation* programs are just two of several programs designed to save wildlife species.

Wildlife Refuges

Wildlife refuges are areas of land and water that provide food, water, shelter, and space for wildlife. The wildlife refuges are set up to save some species from extinction, particularly those on the Endangered Species List. The refuges are maintained, usually by a government or nonprofit organization, for the preservation and protection of one or more wildlife species.

Refer to Volume II for more information about wildlife refuges.

The United States Wildlife Refuge System

The United States Wildlife Refuge has the world's largest, most diverse collection of lands set aside specifically for wildlife. The wildlife refuge system was initiated in 1903 when President Theodore Roosevelt designated Florida's Pelican Island as a *sanctuary* for pelicans and herons. Today 500 national wildlife refuges have been established, ranging in location from the Arctic Ocean to the South Pacific and from Maine to the Caribbean. Most refuges were created to protect the more than 800 species of migratory birds that travel along the four major north-south flyways. The United States has responsibilities under international treaties with Canada, Mexico, Japan, and Russia for migratory bird conservation.

As of 1992, the United States National Wildlife Refuge System, administered by the U.S. Fish and Wildlife Service, comprised some 503 areas, covering 36.5 million hectares (90 million acres) among the 50 states. United States refuges range in size from Minnesota's Mille Lacs (less than an acre/hectare) to Alaska's Yukon Delta at about 9 million hectares (20 million acres). The vast majority of these lands are located in Alaska; the rest are spread across the rest of the United States and several U.S. territories.

Through the Partners for Wildlife Program, the Fish and Wildlife Service provides technical and financial assistance to private landowners who want to restore wildlife habitat on their properties, primarily wetlands, riparian habitat, and native prairie. To date, nearly 11,000 landowners have participated in Partners for Wildlife, restoring thousands of hectares of habitat.

International Reserves

There are many international refuges and preserves throughout the world. The United Nations has established protected areas in many countries called Biosphere Reserves. The Biosphere Reserves are designed to conserve the diversity of plants, animals, and microorganisms that make up the living biosphere. As of 1996, there were 337 biosphere

TABLE 4-3 National Wildlife Refuges Established for Endangered Species

State	Unit Name	Species of Concern	Unit Acreage
Alabama	Blowing Wind Cave NWR	Indiana Bat, Gray Bat	264
Arizona	Buenos Aires NWR	Masked Bobwhite Quail	116,585
Arkansas	Logan Cave NWR	Cave Crayfish, Gray Bat, Indiana Bat, Ozark Cavefish	124
California	Bitter Creek NWR	California Condor	14,054
Florida	Archie Carr NWR	Loggerhead Sea Turtle, Green Sea Turtle	29
Hawaii	Hakalau Forest NWR	'O'u, Hawaiian Hawk	32,730
Iowa	Driftless Area NWR	Iowa Pleistocene Snail	521
Massachusetts	Massasoit NWR	Plymouth Red-bellied Turtle	184
Missouri	Ozark Cavefish NWR	Ozark Cavefish	42
Nevada	Ash Meadows NWR	Devil's Hole Pupfish, Warm Springs Pupfish	13,268
Oklahoma	Ozark Plateau NWR	Ozark Big-eared Bat, Gray Bat	2,208
Texas	Attwater Prairie Checken NWR	Attwater's Greater Prairie Chicken	8,007
Virgin Islands	Gray Cay NWR	St. Croix Ground Lizard	14
Washington	Julia Butler Hansen Refuge for Columbian White-tailed Deer	Columbian White-tailed Deer	2,777
Wyoming	Mortenson Lake NWR	Wyoming Toad	1,776

reserves located in 85 countries, with a total area of 219,891,487 hectares (547,000,000 acres), of which 47 reserves were located in the United States.

One of the largest biosphere reserves is the Serengeti National Park located in Tanzania, in southeast Africa. The park is located in the vast subtropical grassland that is home for more than two million wildebeest, half a million Thomson's gazelles, a quarter of a million zebras, and nearly 500 species of birds.

Another international reserve is located in Costa Rica. This reserve, called the Carara Biological Reserve, was established as part of the country's national park system to help preserve biodiversity. The Carara Biological Reserve, in the Puntarenas Province near the Pacific coast, contains approximately 4,700 hectares (11,614 acres) of land. The reserve is primarily an *ecotone* between the humid rain forest to the south and the dry forest region to the north.

Reserves are regions rich in natural resources. In the Carara Biological Reserve, the primary resources are its flora and fauna, many

Lions are the major predators in the Serengeti National Park. (Courtesy of Jane Mongillo)

Building Wildlife Corridors

Many biologists believe that creating large wildlife preserves in some places in the world is an ineffective preservation strategy. They do agree, however, that it makes sense to connect smaller, fragmented parks and wilderness reserves with wildlife passageways, called wildlife corridors. Today many parks have become isolated from one another by human activities including road building and housing developments.

The wildlife corridor is a migration pathway which connects several isolated, smaller parks with stretches of protective land for animals to roam freely between the parks. The passageways would allow the free movement of wide-ranging animals, such as bobcats, wolves, or lions, to wander from one protective park or preserve to another in search of food and mates. Corridors between the parks have been established by retrofitting highways with underpasses and culverts. For example, a culvert underneath a highway in Santa Ana, California, has been formed into a safe passageway for mountain lions to move from one wilderness preserve to another without much contact with humans. These corridors offer animals protection from crossing busy highways or roaming into densely, human-populated areas where they could be injured or killed.

species of which are almost exclusive to this ecosystem. For example, the area provides habitat to the scarlet macaw, a rare member of the parrot family. The reserve also provides habitat to various monkeys, sloths, armadillos, and iguanas. The Tarcoles River, which borders the reserve, provides a home for crocodiles.

The Carara Biological Reserve has become a popular area for *ecotourism*. During the 1990s, the park had more than 26,000 visitors annually. Despite its protected status, however, human encroachment along the reserve's borders threatens the habitat of some species the park is trying to protect. Illegal mining activities and periodic forest fires also threaten the survival of some park species.

Refer to Volume II for information about captive propagation and national and international reserves.

World Wildlife Fund

Several nongovernment organizations (NGOs) have recognized the need to protect endangered and threatened wildlife species. One of the largest private international conservation organizations is the World Wildlife Fund (WWF). It was founded in 1961 and has more than 50 national chapters across several continents. The WWF, recognized globally by its panda logo, is dedicated to protecting the world's

FIGURE 4-7 • Since 1970, the world rhino population has declined by 90 percent, with five species remaining today—black, white (right), Indian, Javan, and Sumatra—all of which are endangered species.

FIGURE 4-8 • The United States Fish and Wildlife Service removed the bald eagle from the threatened list in 2004. Today, the population of bald eagles are now on the rise in the lower 48 states and may soon be taken off the list entirely.

TABLE 4-4 Bald Eagles at a Glance

Scientific name	*Haliaeetus leucocephalus*
Status	Threatened as of 1994; taken off the list in 2004
Population size	94,500 in the wild (most in northwestern United States and Canada)
Wingspan	2.4 m (8 ft)
Mass	6.8 kg (15 lbs)
Lifespan	40 years in wild; more in captivity

wildlife and their habitats. The WWF has sponsored more than 2,000 projects in more than 100 countries. Its global goals are to protect endangered spaces, save endangered species, and address global threats. The WWF has worked to save the giant panda, tiger, rhinoceros, elephant, whale, and other endangered species. Scientists have identified 200 outstanding terrestrial, freshwater, and marine habitats, such as coral reefs, that need protecting. These ecoregions have been named the Global 200. Global campaigns include species at risk, toxic chemicals, global warming, forests for life, and endangered seas.

Refer to Volume IV for more information on marine habitats such as coral reefs.

In the United States, the Nature Conservancy (TNC) is another NGO dedicated to the preservation of the nation's biodiversity. TNC has acquired over 3.64 million hectares (8.5 million acres) of wildlife habitat and manages over 1,500 reserves. Currently focusing on developing agreements with the business community, TNC has come to an agreement with the timber company Westvaco to conduct a biodiversity inventory of its 562,000 hectares (1,350,000 acres) of forested land.

Vocabulary

Captive propagation The controlled mating and breeding of captive animals and plants in zoos, aquariums, botanical gardens, and private research institutions.

Clear-cutting Removing all trees from aforested area.

Deforestation Cutting down trees for commercial use primarily by clear-cutting techniques.

Ecotone An area between two different types of vegetation, such as the border between a desert and a grassland.

Ecotourism A segment of the travel industry that centers around people visiting places known for their unique natural beauty or their exotic or abundant wildlife.

Exploitation To profit from the use of labor or materials in an unethical way.

Sanctuary Special area where wildlife are protected.

Siltation The action of depositing silt at the bottom of water.

Watershed The area of land that drains into a lake, river, or stream; a boundary or line between the headstreams of river systems; a basin divide between headstreams of a river.

Activities for Students

1. Make a list of products that you and your family use which could have come from a rain forest. How can we as consumers be sure that they were produced in an environmentally sustainable fashion?
2. Create an informational advertising campaign, in video or print, which stresses the importance of supporting sustainable forest use.
3. Pick one animal from the list of endangered species. Research its habitats, how it is threatened, and how it is being protected.
4. Plan an imaginary ecotourism trip to an international reserve. Where would you go, and how would you be respectful of the natural surroundings? Find out how to get there, where to stay, and what to do while you are there.

Books and Other Reading Materials

Ashabranner, Brent. *Morning Star, Black Sun: The Northern Cheyenne Indians and America's Energy Crisis.* New York: Dodd, Mead, 1982.

Breymeyer, A. I., D. O. Hall, J. M. Melillo, *Global Change: Effects on Coniferous Forests and Grasslands.* Scope, no. 56. New York: John Wiley, 1997.

McClung, Robert. *Lost Wild America: The Story of Our Extinct and Vanishing Wildlife.* Hamden, Conn.: Linnet Books, 1993.

O'Brien, Bob. *Our National Parks and the Search for Sustainability.* Austin: University of Texas Press, 1999.

Sayre, April Pulley. *Temperate Deciduous Forest (Exploring Earth's Biomes).* New York: Twenty First Century Books, 1995.

Seymour, John, and Herbert Girardet. *Blueprint for a Green Planet: Your Practical Guide to Restoring the World's Environment.* Upper Saddle River, N.J.: Prentice Hall, 1987.

Terborgh, John. *Diversity and the Tropical Rain Forest* Scientific American Library, no. 38. New York: W. H. Freeman, 1992.

Websites

Defenders of Wildlife, www.defenders.org

Eco World: NGO supporting reforesting activities in both urban and wilderness environments around the world through education and direct planting operations, www.ecoworld.org

Fauna and Flora International, which acts to conserve threatened species and ecosystems worldwide, choosing solutions that are sustainable and are based on sound science, http://www.fauna-flora.org/

National Forest Foundation, http://www.nffweb.org

National Wildlife Federation, http://www.nwf.org/

Site listing 140 species under threat, http://www.wcmc.org.uk/species/data/species_sheets/

Society of American Foresters, http://www.safnet.org

Trees, Water & People: NGO working cooperatively with communities to establish and maintain sustainable forests, watersheds, and wetlands while improving people's lives, www.treeswaterpeople.org

U.S. Fish and Wildlife Service, listing of endangered and threatened species, http://endangered.fws.gov/wildlife.html#Species

U.S. Forest Service, *Roadless Area Review and Evaluation*, which evaluates lands as "wilderness areas," http://www.fs.fed.us

U.S. National Park Service, http://www.nps.gov/

World Conservation Monitoring Centre, http://www.wcmc.org.uk

World Conservation Union, http://www.iucn.org/about/index.htm

World Conservation Union Red List, http://www.redlist.org/

World Resources Institute Forest Frontiers Initiative, http://www.wri.org/ffi

World Wildlife Fund, http://www.worldwildlife.org

CHAPTER 5

Sustainable Business Stewardship

Hundreds of nations are currently exporting and importing large quantities of products. These products, being traded back and forth every day, include everything from automobiles, television sets, and computers to sports equipment, clothing, and food products. This widespread *economic growth* has brought social progress to many countries in the world. Food products are now more accessible and affordable, literacy rates have improved, and life expectancy has increased. A strong and growing economy has provided a higher standard of living for many people.

Economic growth and progress, however, come at a price. In the last 40 years or so, the enormous global economic and population growth has caused environmental and social issues ranging from air and water pollution and the degradation of natural resources to human rights issues and damage to human health. Many agree that the current patterns of consumption and production of businesses, manufacturing, and other industries' are not very sustainable. With all of these environmental problems, is it possible to have a fair and just economy which incorporates sustainable business practices?

BUSINESS STEWARDSHIP: ECO-EFFICIENCY

Many progressive business leaders, government groups, and public interest groups are looking for sustainable business practices that balance ecological, economic, and social goals. The concept of merging or integrating ecological, social, and economic goals is called eco-efficiency. The term was introduced in 1992 by the World Business Council for Sustainable Development (WBCSD). They described eco-efficiency as a means of producing economically valuable *goods* and *services* while reducing the ecological impacts of production.

Refer to Volume III for more information about goods and services.

The WBCSD provides a business leadership that promotes eco-efficiency through high standards of environmental and resource management in business. The WBCSD comprises member companies from 34 nations throughout the world, including Xerox Corporation, DuPont Corporation, and General Motors.

Goals of Eco-Efficiency

The WBCSD has identified several goals of eco-efficiency that every business should take into account when developing products:

- Reducing the amount of material used in manufacturing goods
- Becoming less dependent on fossil fuel energy
- Promoting recyclability of goods
- Reducing toxic emissions and other pollutants
- Using more sustainable resources in the production of goods

By implementing these goals, businesses will produce more and better products from the same amount of raw materials with less waste and fewer adverse environmental impacts. Using eco-efficiency guidelines means that businesses will need to monitor and evaluate the environmental impact at every stage in the production of goods. Some critics oppose the use of the definition of eco-efficiency in which industry uses less energy and produces less waste. They argue that "less bad" is not correct. Industries should be more concerned that they are doing the right thing.

ECO-EFFICIENCY AND THE LIFE CYCLE OF A PRODUCT

The vision of eco-efficiency is to make the manufacturer responsible for products throughout the production stages of the life cycle. A typical life cycle of a product includes the extraction of raw materials, the conversion of raw materials into manufactured goods, the marketing and servicing of goods, the consumption of goods, and the disposal of them. In each stage, there is a relationship between the materials used in the production and its impact on natural resources and energy use. All of these stages require natural resources, human resources, *capital resources*, and energy. At each stage in the life cycle a certain amount of environmental wastes and pollution is generated. The challenge of initiating eco-efficiency practices, therefore, is to reduce the waste at each stage using less energy and materials while producing a valuable good or service.

One way to reduce waste during the life cycle of a product is to separate waste into recyclable and nonrecyclable uses. This process, called source separation, makes recycling more efficient because it takes place right at the source where the waste is produced. Source separation removes all designated recyclable materials from the waste stream and makes it available to be reused. For example, printers use special paper cutters to trim pages in book and magazines. The leftover waste, at the source, is placed in bins to be recycled rather than to be tossed out and carried eventually into landfills. Many books and magazines are printed on recycled paper.

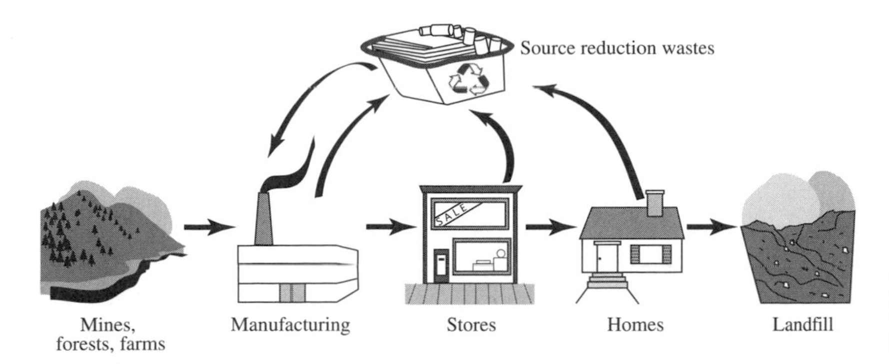

FIGURE 5-1 • System of Production and Consumption

Eco-Efficiency in Manufacturing Automobiles

Sustainable economists agree that cars should not be built to last forever because eventually we would be overpopulated with cars. Instead, the materials of a discarded car could infinitely be recycled into new ones. Several automobile manufacturers are currently developing ecoefficiency plans for recycling and disposing of automobiles that are no longer usable. For example, many different kinds of plastics are used in making automobiles. Plastics are used for seats, the trim around windows and doors, dashboards, cupholders, engine parts, and bumpers. When the car is disposed of, the different plastic parts are mixed together making it difficult and costly to recycle and reprocess them. Some automakers, including Volkswagen, have made a change to help solve that problem. Volkswagen now marks all of the different plastics that are used in the car's construction so that someday, when it is demolished, it will be easy to identify the plastics and to sort them out appropriately. Much of the plastic in the old Volkswagens can then be recycled and reused.

The dismantling of products where parts are removed and recycled in the automobile industry is not new. In the 1930s, Henry Ford recalled hundreds of old Fords to his automobile assembly line to disassemble the metal parts for recycling. Ford also used old, discarded shipping crates for wood parts in the construction of automobiles.

Henry Ford posing beside the first and the ten millionth Ford. Ford disassembled wood and metal parts from old Fords and recycled the materials for building new autos. (Courtesy of the Library of Congress)

Some Eco-Efficiency Companies

How many businesses have initiated eco-efficiency business practices? According to the World Resource Institute and other business experts, several progressive businesses, mostly large ones, have adopted ecoefficiency practices.

Proctor and Gamble

In 1989 Proctor and Gamble introduced a new kind of detergent powder, called Ultra detergent. The packaging of the product took up half the volume of that of traditional detergents but cleaned the same amount of clothes. The smaller package was more convenient for consumers to handle. The product used 30 percent fewer raw materials and 30 percent less packaging. Furthermore, shipping the smaller product reduced the amount of energy needed to transport it to the marketplace.

3M Corporation

An early pioneer, 3M Corporation now claims that its Pollution Prevention Pays program has prevented more than 750,000 metric tons of polluting emissions since 1975 by cleaning up and redesigning processes and products while saving the company hundreds of millions of dollars.

SC Johnson Wax

An eco-efficiency program established at SC Johnson Wax in 1990 has cut the company's manufacturing waste by half and has reduced packaging waste by 25 percent. The company also reduced the use of volatile organic compounds by 16 percent; at the same time, production has increased by more than 50 percent. The company's largest plant extracts methane gas from a nearby landfill to meet a substantial portion of its energy needs; another plant continuously reuses 95 percent of its wastewater so that it is never discharged. These innovations have saved the company millions of dollars in annual costs.

Xerox

In some cases, as with office equipment, manufacturers are turning to reconditioning or rebuilding old equipment, rather than building every new machine from scratch. Xerox, for example, has developed a product return practice to reclaim old copiers for reconditioning. The company has found that even recycling low-value items such as toner cartridges can be profitable. In 1994 Xerox saved $2 million in raw material costs by reusing toner cartridges—enough to cover the costs of collecting the cartridges, including a cash incentive program to encourage customers to join the recycling effort.

Rechargeable Battery Recycling Corporation

The Charge Up to Recycle program, initiated by the Rechargeable Battery Recycling Corporation, educates the public about the need to recycle nickel-cadmium (Ni-Cad) batteries, which are a major source of toxic metals found in solid waste landfills. The program is funded by more than 20 companies that manufacture rechargeable batteries for

Used power tool rechargeable batteries can be deposited in special recycling bins at The Home Depot and other businesses. The Rechargeable Battery Recycling Corporation sponsors the program. (Courtesy of The Rechargeable Battery Recycling Corporation [RBRC]. Log on to the following Website to learn more about this program: http://www.rbrc.org/)

TABLE 5-1 Types of Batteries Accepted by INMETCO Recycling Facility

Battery Systems	Applications
Lithium ion	Computers, portable phones
Lead acid, automobile	Automobiles, tractors, marine
Magnesium	Transmitters, aircraft transmitters
Mercury	Hearing aids, military sensors and detection tools
Nickel cadmium, industrial	Railroads, communications, aircraft, utilities
Nickel cadmium, dry-sealed	Portable phones, tools, computers, appliances
Nickel iron, industrial	Railroad signals
Nickel metal hydride	Portable phones, computers
Zinc carbon	Flashlights, toys radios, instruments

INMETCO, a company located in Pennsylvania, recycles the nickel, chrome, and iron in the byproducts of stainless steel production so that the materials can be reclaimed and reused instead of landfilled. The company also recycles a large variety of batteries.

North American sales. Batteries, collected from businesses, government agencies, and consumers, are recycled. The Ni-Cad batteries do not end up as trash in landfills or incinerators.

Interface Flooring Systems

The Evergreen program, sponsored by Interface Flooring Systems, is a new way to sell carpet. In the program, the consumer leases the services of replaceable carpet tiles without having to dispose of the whole carpet when it wears out. Instead of throwing out the entire carpet when one section wears out, the consumer replaces the worn-out area with individual tiles, as needed.

Some Examples of Sustainable Products

Several products have received high marks for eco-efficiency and sustainability:

- Locally grown, organic food and drink products
- Clothing made from Foxfibre cotton, which is organically grown and naturally colored
- Solar water heaters, cookers, and other appliances
- Bicycles made from repairable and recyclable materials including parts made from steel, aluminum, and rubber
- Packaging made from gourds and other natural materials
- Furniture using wood, leather, and other materials that can be disassembled and remanufactured.

Eco-Industrial Parks

Another unique idea implementing the concept of eco-efficiency is to develop centers called eco-industrial parks. A typical eco-industrial park houses a group of different kinds of businesses that work together with the local community in sharing resources and in improving the environment. Some joint projects include activities to reduce emissions and wastes, to generate electricity by using biomass technology, to recycle and compost local waste products to replenish soils, and to protect local wildlife.

Different types of eco-industrial parks have been established in Texas, Vermont, Virginia, and other states. The Riverside Eco Park in Burlington, Vermont, a proposed agricultural-industrial park, will be located in an urban area. The plans are to use the waste product, wood chips, from one of its businesses to generate electricity. The waste heat would be used by a local greenhouse to raise marketable fish and plants. The center will also use an ecological system to break down and digest liquid waste to purify water and produce high-grade fertilizers. Establishing eco-industrial parks in urban areas can also provide local job opportunities in cities which have a high poverty rate and high unemployment rate.

Eco-Efficiency in School

A number of schools have adapted some of the business world's eco-efficiency goals into their own school activities. The schools have found ways to reduce energy use, cut down on waste, and recycle materials. The following suggestions can be used by students to help:

- Research and use a wide assortment of products made from recycled products, such as pencils made from old blue jeans, binders made from old shipping boxes, and many types of recycled paper products.
- Before starting a new school year, sort through your materials. Many supplies can be reused or recycled. Notebooks, folders, and binders can be reused.
- Waste from packaging accounts for more than 30 percent of all the waste generated each year. Use school supplies wrapped with minimal packaging, use compact or concentrated products, or buy products that come in bulk sizes.
- Many schools reuse textbooks to save money and reduce waste. Covering your textbooks with cut-up grocery or shopping bags helps reduce waste and keeps your books in good condition. Be creative—use markers or colored pencils to give your covers unique and fun designs. Paper grocery bags are also great for wrapping packages.
- Use nontoxic products, inks, and art supplies, such as batteries with less mercury, vegetable-based inks, and water-based paints.
- Use and maintain durable products. Sturdy backpacks and notebooks can be reused for many years, which helps reduce the number of unusable items tossed away each year.
- If you bring your lunch to school, package it in reusable containers instead of disposable ones, and carry them in a reusable plastic or cloth bag, or lunch box. Bring drinks in a thermos instead of disposable bottles or cartons.
- If you buy lunch, take and use only what you need: one napkin, one ketchup packet, one salt packet, one pepper packet, one set of flatware. Recycle your cans and bottles.
- Volunteer for an environmental club or recycling project in your school.

BUSINESS STEWARDSHIP: ECO-LABELING

Government organizations, consumer groups, and others are now labeling a wide range of goods that have been produced by manufacturers, farms, and fisheries which practice sustainability. The labels on the goods, commonly known as eco-labels, promote natural resources and energy conservation, human labor rights, fair trade, and agricultural goods free from pesticides and fossil fuel–based fertilizers.

Examples of Eco-Labels

Eco-labeling, which is now used in many countries, was first used in Germany in 1977. Germany introduced a consumer program called the Blue Angel. The Blue Angel's environmentally friendly label is placed on food packages to show the item's environmental benefits. For example, an eco-label placed on a package could indicate that the product is recyclable or biodegradable, or it was sustainably produced.

The Blue Angel label encourages environmental awareness, promotes natural resource conservation, and limits pollution. The Blue Angel label has currently certified 4,000 products for German consumers.

In 1982 the U.S. Federal Trade Commission set up guidelines for the use of the labels "recyclable," "compostable," and "biodegradable." Other countries, such as Canada, Japan, and India, have environmental labeling certifications for the products that pass government tests for environmental benefits such as recyclability.

A number of consumer groups, government agencies, and environmental organizations have developed their own eco-labeling programs to evaluate and certify products that promote sustainable practices. By buying certified products, such as refrigerators, paper goods, groceries, fish, and furniture, the consumer is supporting sustainable farming and forestry practices, limiting pollution, and reducing dependency on fossil fuels. Some of the eco-labels seen in stores are discussed below.

Rainforest Alliance Certified

The Rainforest Alliance label is used to certify bananas, coffee, and orange juice produced from beans and fruit grown on farms promoting sustainable agricultural practices. Their seal of approval appears on food grown in Central and South America, the Philippines, and Hawaii.

Salmon Safe

The Salmon Safe label appears in many natural food stores on a variety of products including wine, fruit, milk, and rice. The organization promotes farms that conserve soil and preserve wetlands and streams that benefit nearby native salmon fisheries in the Pacific Northwest.

Green Seal

The Green Seal label, the oldest U.S. eco-label, started in 1989, is used to certify about 300 products ranging from paper and cleaning supplies to sustainable alternative-fuel vehicles.

Green-e

The Green-e label certifies green-power companies which use renewable energy sources such as solar, wind, geothermal, biomass, and small hydro-powered plants. Green-e residential and commercial users are nationwide.

Energy Star

This label, probably the most commonly seen label in the United States, appears on some 11,000 products including energy-efficient appliances, building supplies, heating and cooling equipment, and stereo systems.

Marine Stewardship Council

The Marine Stewardship Council (MSC) certifies sustainably managed fisheries. To be certified, the fishery must use fishing practices that do not take more fish than can be replenished naturally, ensure the health of the marine ecosystem, and respect local, national, and international laws for responsible and sustainable fishing. The MSC label lets consumers know that when they buy MSC-labeled seafood, they are supporting healthier oceans and a healthier environment.

Be Cautious of Eco-Labels

Some eco-labeling has come under fire by consumer groups and environmentalists because of false claims and other abuses. They state that some products, require more research and testing before it can be determined whether one product is less harmful to the environment than another.

The consumer needs to be well informed and responsible in using products with eco-labels. Some companies use eco-labels as a marketing gimmick and have vague product standards and descriptions. Consumer groups advise shoppers to do their research before they buy any product with an eco-label. Eco-labeling is a good idea, but consumers need to change their lifestyles by being more aware of the products that they are buying and using and how these products impact the environment. Eco-labeling organizations can be researched on the Internet.

ACCOUNTING FOR ECO-EFFICIENCY AND SUSTAINABILITY

Gross Domestic Product

Several businesses have made major strides in developing sustainable and eco-efficient practices, but many of them believe that their governments should help them do more. At the Earth Summit held in Rio de Janeiro in 1992, the WBCSD urged governments to tax energy and to stop subsidizing fossil fuels. Members also proposed measures to incorporate environmental accounting in the *gross domestic product* (GDP). The GDP is the tool most widely used to measure the success of a country's economy or wealth. The GDP measures the total market value of the country's goods or services produced per capita by labor and property during a year.

By knowing a country's GDP per capita, one can understand a country's social, economic, and environmental strengths by measuring the general standard of living enjoyed by the inhabitants. Populations in countries with a high GDP tend to have longer life expectancies, better health, higher education, better access to clean water and air, and lower infant mortality than those in the middle or at the bottom levels of GDP.

DID YOU KNOW?

About 20 percent of the world's population in the industrialized developed countries has 80 percent of the world's GDP.

Countries with the highest incomes per capita use much of the natural and energy resources, such as fossil fuels, and cause the majority of global environmental problems, such as water and air pollution.

Another criticism of using the GDP measuring tool comes from environmental economists. They contend that the GDP does not take into account the losses and depreciation of natural resources used for mining, fishing, logging, industrial, and farming activities or industrial uses. For example, the accounting in a GDP statement would include a country's sale of wheat minus the costs of the depreciation of farm equipment and buildings used to grow and store the wheat. However, the GDP does not adjust for any loss of natural resources such as the

TABLE 5-2 Gross Domestic Product per Capita, 2000

Some Countries above $9,361		Some Countries between $761 and 9,360		Some of the Countries at $760 or less	
United States	$33,900	Saudi Arabia	$9,000	Cambodia	$710
Singapore	$27,800	Poland	$7,200	Ethiopia	$560
Japan	$23,400	Brazil	$6,150	Tanzania	$550
France	$23,300	Russia	$4,200	Sierra Leone	$500
United Kingdom	$21,800	China	$3,800		
Chile	$12,400	Egypt	$3,000		
South Korea	$13,300	India	$1,800		

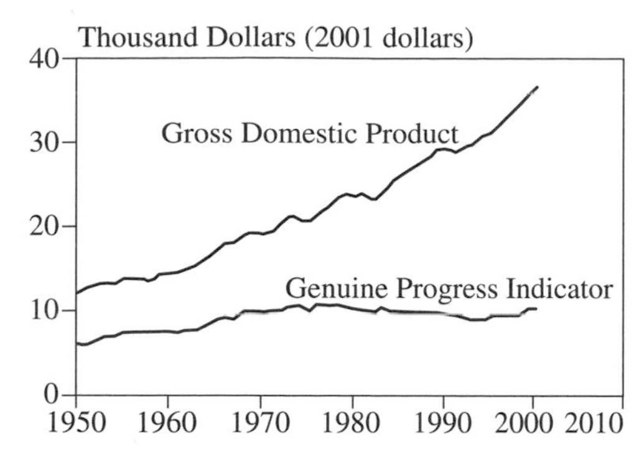

FIGURE 5-2 • Gross Domestic Product and Genuine Progress Indicator per Person, United States, 1950–2001 The Genuine Progress Indicator (GPI) was created by Redefining Progress, a U.S. nongovernment research organization. The GPI subtracts costs to the economy such as crime, pollution, and traffic. *Source:* Redefining Progress Organization.

amount of water used to grow the crops or the depletion of fertile soil caused by soil erosion or the costs of pesticides. Nor does the GDP take in account other debits or expenditures such as the cleanup of air and water pollution. Cleanup costs often fail to take into account cumulative or long-term environmental damage. In fact, when cleanup costs are acquired by a community, government, or business, the GDP increases; for example, the costs of cleaning up an oil spill adds to the GDP, a negative expenditure.

DID YOU KNOW?

The GDP of Alaska increased by one billion dollars in 1989 because of the Valdez oil spill.

Teen-Age Projects on a Sustainable Economy

In 1998 high school students in Washtenaw County, Michigan, collaborated with the county government to present a public forum on sustainability. Informative presentations, skits, and music brought to life global and regional development issues for an audience of over 100 parents and community leaders. The students' projects, developed in their civics class, ranged in topic from local agriculture to global trade.

In the skit "Oil Spills are Good for Us!," a student representing the GDP applauded oils spills and other environmental problems because their cleanup adds to our nation's bottom line. The performance illustrated the disturbing point that pollution, divorce, and crime are good economic news because their impacts contribute to the GDP. In a presentation following the skit, the students proposed new indicators that would account for the social and environmental effects of transactions. "Imagine if we subtracted the costs of environmental degradation and social decay from the gross domestic product," they noted. "What would it look like then? Would we still seem to be making progress and headed in the right direction?"

For the presentation "What's the Real Cost of Driving? Who Pays?," the students presented data on the external costs of car-using data—the social and environmental impacts not reflected in the market price. Their figures, based on research from the think tank Redefining Progress, included the following:

- Accidents: $77.5 billion
- Loss of farm land: $80.5 billion

(*continued*)

Teen-Age Projects on a Sustainable Economy (*continued*)

- Depletion of nonrenewable resources: $793.5 billion
- Ozone: $213.8 billion
- Air pollution: $34.6 billion.

The students then explained how these costs are borne by the public and suggested economic "carrots" and "sticks" that would promote sustainability through more accurate prices.

The curriculum and instruction for the project were provided by Creative Change Educational Solutions, a nonprofit sustainability education organization. For over six weeks, Creative Change staff worked in the classroom alongside the teacher to deliver lessons, facilitate activities, and support the student-directed projects. The basis of the program was Creative Change's ecological economics curricula *The Shape of Change.*

The innovative partnership was supported by Washtenaw County government as part of its continuing efforts to promote regional sustainability. Creative Change has provided programming and curricula to other districts in southeast Michigan, including the Ypsilanti public schools. In a project funded by the U.S. Environmental Protection Agency, Creative Change provided professional development to help Ypsilanti teachers integrate sustainability into language arts, social studies, science, and math courses.

Further information on Creative Change is available at www.creativechange.net. Photos and text by Susan Santone.

Freshman students at Ypsilanti High School explore the social and environmental impacts of local development trends. The interdisciplinary unit included a regional tour of land use patterns, a trip to the solid waste processing facility, research on community history, and peer presentations of their findings. (Courtesy of Susan Santone/Creative Change)

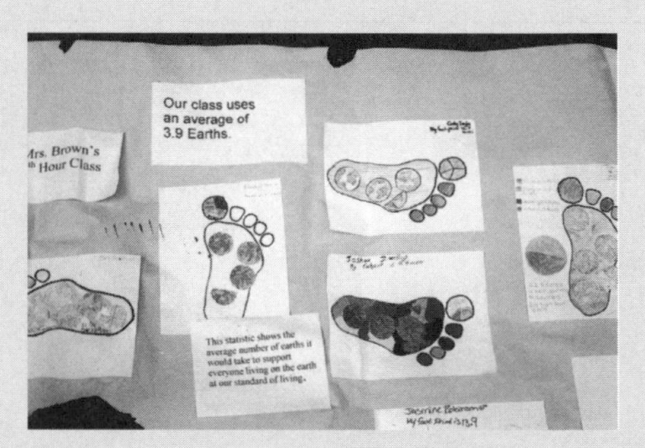

Eighth graders at Ypsilanti's West Middle School explored the question: *How could population and consumption affect the future?* To examine their personal impact on the environment, students used an online program to calculate their "Ecological Footprint," which is the amount of resources needed to support their lifestyle. See http://www.myfootprint.org for more information. (Courtesy of Susan Santone/Creative Change)

The GDP per capita does not indicate whether all people share in the wealth of a nation nor does it measure the quality of life of people or whether the people are leading fulfilling lives. As a result of this criticism of using the GDP as an indicator of a country's wealth, several economists are developing alternative measuring tools. Many of

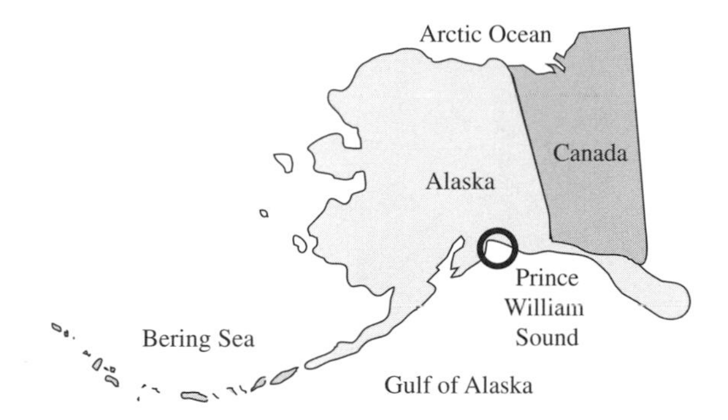

FIGURE 5-3 • Gulf of Alaska The Exxon *Valdez* oil spill was one of the most environmentally damaging oil spills in history. The Exxon *Valdez* oil spill occurred in Prince William Sound, off the coast of Alaska, on March 24, 1989.

TABLE 5-3 Professions in the Eco-Economy

Wind meteorologists	Recycling engineers
Hydrologists	Ecological economists
Aquacultural veterinarians	Geothermal geologists
Bicycle mechanics	Wind turbine engineers
Foresters	

these plans include adjusting the GDP to make it a more accurate representation of the good that people receive from their nation's economic growth. One goal is to subtract indirect costs from the GDP ledger. Indirect costs are those costs associated with the destruction of natural resources and the cost of medical care for victims of pollution, particularly in urban areas.

Environmental economists also argue that economic growth does not necessarily contribute to a person's well-being if it diminishes Earth's natural resources. Besides focusing on the reduction of energy use, product waste, recycling and reusing materials, the economic plan must get input from local groups to help determine which economic system is best for them. Other goals in an eco-economy plan include creating jobs, establishing local job training, encouraging educational opportunities, and providing fair and just wages without increasing material consumption. Whether these goals will be attained is the question. The answer lies with individuals, governments, and business corporations, who need to work together to adopt new measures for a sustainable economic future.

Vocabulary

Capital resources Funds or other assets contributed to a business by the owners or stockholders.

Economic growth Occurs when an economy is able to produce more goods and services over time and is measured in gross domestic product.

Goods Any physical object, such as a car, that satisfies a person's wants or needs.

Gross domestic product Total dollar value of all newly produced goods and services in an economy in a given year.

Services Any productive activity, that satisfies a person's wants or needs.

Activities for Students

1. Find 10 products in your house and make a list of where they come from. What countries produce many of the goods we use, and why do you think that they have large manufacturing industries in those places? What kind of impact might that have on the environments of those countries?
2. Pick one complex item or product, such as a toy, and track its life cycle. Think about what raw materials it contains, how it was produced, and what happens to it after it is discarded.
3. Create a model of an eco-industrial park. Consider what features it would have that would make it an attractive place for a large corporation to do business.
4. Visit a local food store and look for products with eco-labels. What kinds of products generally carry these labels, and what impact may that have on the consumers making choices about what items to buy?

Books and Other Reading Materials

Brown, Lester R. *Eco-Economy, Building an Economy for the Earth.* New York: W. W. Norton, 2001.

Carless, Jennifer. *Taking Out the Trash: A No-Nonsense Guide to Recycling.* Washington, D.C.: Island Press, 1992.

Davidson, Eric A. *You Can't Eat GNP: Economics as if Ecology Mattered.* Cambridge, Mass.: Perseus Publishing, 2000.

Foster, Joanna. *Cartons, Cans, and Orange Peels: Where Does Your Garbage Go?* New York: Clarion Books, 1991.

Hawken, Paul. *The Ecology of Commerce: A Declaration of Sustainability.* New York: HarperCollins, 1993.

Henderson, Hazel, Jon Lickerman, and Patrice Flynn, *Calvert-Henderson Quality of Life Indicators.* Bethesda, Md.: Calvert Group, 2000.

Kalman, Bobbie. *Reducing, Reusing, and Recycling.* New York: Crabtree Publishing, 1994.

McDonough, William, and Michael Braungart. *Cradle to Grave: Remaking the Way We Make Things.* New York: North Point Press, 2002.

Rifkin, Jeremy. *The Hydrogen Economy: When There Is Not More Oil.* New York: Tarcher/Putnam, 2002.

Silverstein, Alvin. *Recycling: Meeting the Challenge of the Trash Crisis.* New York: G. P. Putnam, 1992.

Websites

CERES, http://www.ceres.org or e-mail ceres@igc.apc.org

Green Seal, http://www.greenseal.org

Recycling information, www.epa.gov/kids/garbage.htm; http://www.epa.gov/recyclecity/

Smart Office, http://www.smartoffice.com/

U.S. Environmental Protection Agency (from ecosystems to conservation, human health to waste and recycling, air to water; includes instructions on internships, interactive games on environmental topics, and various club projects for viewers to browse), http://www.epa.gov/students

U.S. Environmental Protection Agency, http://www.epa.gov/epaoswer/osw/specials/funfacts/school.htm

Wisconsin Department of Natural Resources (process of recycling; viewers interested in all the ins and outs of recycling, how to be a vermiculturist, composting with worms, and recipes for composting will enjoy this site), http://www.dnr.state.wi.us/org/caer/ce/eek/earth/recycle/index.htm

CHAPTER 6

Sustainable Communities and Transportation

In the 1960s, only three cities in the world had a population of more than 8 million inhabitants: New York City; Tokyo, Japan; and London, England. In the year 2015, population experts predict that as many as 30 cities throughout the world will exceed a population of 8 million. By 2007 more than 50 percent of the world's population will live in cities, and by 2050 almost 7 out of every 10 people in the world will be city inhabitants.

The big jump in population growth is causing *urban* sprawl. Urban sprawl occurs when the edges of a city spread outward into the adjoining suburbs and rural sections. For the most part, sprawl fans out in a continuing, disorderly, chaotic fashion which seems unstoppable. Looking down from an aerial view, one sees an urban-suburban sprawling network of shopping malls, industrial parks, and the construction of new housing developments and roads which are all linked together by multilane superhighways.

DID YOU KNOW?

Urban sprawl started right after World War II, when private ownership of cars became common. In the 1950s, the massive development of superhighways allowed workers to live in small towns, where housing was affordable, and commute to work by car.

PROBLEMS CAUSED BY SPRAWL

In the United States, sprawl is growing at a rate of 146 hectares (365 acres) of land per hour, according to government figures. In fact, in most communities, the amount of developed land needed for houses, businesses, and roads is growing faster than the population. In Ohio, for example, the amount of development around urban areas between 1960 and 1990 grew more than five times as fast as the population, according to an article published in *Time* magazine.

Sprawl causes and accelerates a whole list of environmental and social issues such as air and water pollution, the destruction of open spaces and wildlife habitats, and the loss of biodiversity. According to one environmental organization, more than 400,000 hectares (one million acres) of parks, open space, and wetlands are destroyed each year for land development. One of the most serious problems of sprawl is the loss of agricultural land. According to one study, about 560,000 hectares (1.4 million acres) of prime agricultural land is converted to land development each year in the United States. As a

TABLE 6-1 Population of World's 10 Largest Metropolitan Areas in 1000, 1900, and 2000

City	1000 (million)	City	1900 (million)	City	2000 (million)
Cordova	0.45	London	6.5	Tokyo	26.4
Kaifeng	0.40	New York	4.2	Mexico City	18.1
Constantinople	0.30	Paris	3.3	Mumbai (Bombay)	18.1
Angkor	0.20	Berlin	2.7	São Paulo	17.8
Kyoto	0.18	Chicago	1.7	New York	16.6
Cairo	0.14	Vienna	1.7	Lagos	13.4
Bagdad	0.13	Tokyo	1.5	Los Angeles	13.1
Nishapur	0.13	St. Petersburg	1.4	Calcutta	12.9
Hasa	0.11	Manchester	1.4	Shanghai	12.9
Anhilvada	0.10	Philadelphia	1.4	Buenos Aires	12.6

Source: Molly O'Meara Sheehan, *Reinventing Cities for People and the Planet*, Worldwatch Paper 147 (Washington, D.C.: Worldwatch Institute, June 1999), pp. 14–15, with updates from United Nations, *World Urbanization Prospects: The 1999 Revision* (New York: 2000).

result of land development, in the last 20 years, agricultural property available for growing grain has dropped by 7 percent.

In urban areas, sprawl-like conditions result in high traffic levels and congestion. Emissions from motor vehicles, particularly cars, cause air pollution. In fact, 75 percent of all air pollution in a city is caused primarily by car traffic.

Sprawl also costs taxpayers money. Communities need to appropriate funds for the construction and maintenance of roads, bridges, and other *infrastructure* costs. Taxes are raised to pay for new water and sewer lines, new schools, and more fire and police protection.

STRATEGIES TO CONTAIN SPRAWL AND BUILD SUSTAINABLE COMMUNITIES

Citizens, public interest groups, and local and state governments have begun to develop strategies to contain sprawl. The strategies include setting boundaries for growth, providing more parks and adding green-space areas, restoring inner cities, designing sustainable communities, and investing in alternative forms of transportation. As a result of these plans farmland and wildlife areas will be preserved and air and water pollution will be reduced.

Setting Growth Boundaries for Communities and Providing More Green Space

Community planners are using growth management strategies to establish specific land areas for development and other areas to remain undeveloped. Roads, water and sewer lines, and housing would be built only in designated areas. The underdeveloped green-space areas, called *greenbelts*, would be planned in and around the community to protect farmland, preserve local wildlife habitats, and provide open space and recreational areas for local people to enjoy.

Portland, Oregon, is well known for establishing growth management plans to protect open space, wildlife habitats, and farmlands. Since 1979 development has been off-limits in greenbelt areas. By concentrating the population in a denser area, the costs for building roads, houses, and sewer lines is much less than if the city did not set growth boundaries, allowing sprawl.

Providing the greenbelts within a community adds trees and plants which create cool environments, reduce noise levels, and remove pollutants from the air. Greenbelts also make communities more aesthetically attractive. Many countries, such as the Netherlands, France, and Germany, as well as the United States, have established greenbelt programs. Establishing greenbelts encourages developers and architects to be more sensitive to the contour of the land, soil quality, water drainage, and areas of wildlife habitats when planning new housing developments.

Purchasing Land

Buying land is another way to curb sprawl and development. Governments, private organizations, towns and cities, and even individuals have purchased land to keep it off the marketplace. As an example, the founder of a major clothing company bought 256,000 hectares (640,000 acres) of forest land in Chile to protect the land from development.

Each year, the U.S. Congress appropriates money to purchase undeveloped land. Several states, such as New Jersey, have issued bonds for the preservation of local farms and woodlands. In Pittsford, New York, a small suburb of Rochester, bonds were sold to buy 480 hectares (1200 acres) of seven farms which would no longer be put to agricultural uses.

Open space, fields, forests, and wetlands are purchased by conservation groups. As one example, the Nature Conservancy is an international organization dedicated to protect critical wildlife habitats. One way they protect lands is by purchasing the land outright or by acquiring *conservation easements*. A conservation easement gives the rights to an environmental organization to use the land of a private owner, but the land is to be conserved and restrictive on economic uses. The organization is responsible for protecting approximately 1.2 million hectares (3 million acres) of land in the United States, Canada, and South America.

Restoring Cities into More Sustainable Communities

Many cities have suffered from declining neighborhoods, high unemployment, and air and water pollution. These problems have caused city dwellers to move, live, and work outside the city, causing more sprawl. City planners have proven, however, that older towns and cities can be rebuilt, revitalized, and made more hospitable to encourage city dwellers to stay in the city. Chattanooga, Tennessee, is a good example of an urban area that has been revived.

In the 1970s, Chattanooga was a city experiencing economic decline, vacant and crumbling buildings, and serious air pollution. In 1984 city planners, community members, civic leaders, and government agencies began tackling these problems. Since that time, the downtown area has been revitalized, the riverfront area has been cleaned up and

Chattanooga, Tennessee has been rebuilt and revitalized with new parks, a riverwalk, electric buses, and other improvements. (Provided by the Chattanooga Area Chamber of Commerce)

new parks created, low-cost housing has been renovated, and air and water pollution has been reduced. Many city jobs were established, and a zero-emissions industrial park was constructed. A new industry was founded in the city to build nonpolluting electric buses, which are now operating.

Seattle, Washington, was experiencing many of the same problems as Chattanooga. Today, however, the city is ranked as one of the most livable communities in the United States. Because of its popularity as a nice place to live, Seattle city planners expect the population to grow at a rate of more than 20,000 people each year until 2010. A citizen-led project, along with city government, is addressing environmental priorities by creating urban-village centers within Seattle while reducing sprawl in the adjoining suburbs. Part of the plan is to establish a neighborhood of parks with low-income housing and pedestrian-friendly streets.

Cleaning Up Brownfields

Other cities are taking action to clean up environmentally contaminated land, known as *brownfields*, to make them more attractive for redevelopment. The U.S. Environmental Protection Agency (EPA) defines brownfields as "abandoned, idled, or under-used industrial and commercial facilities where expansion or redevelopment is complicated by real or perceived contamination."

Old industrial properties which may have been abandoned after going out of business are often considered brownfields. The soil or groundwater on the brownfield site may contain oil or hazardous materials that would be costly to clean up or could present a threat of harm to health, safety, or the environment. When industrial properties are abandoned, the city government cannot obtain taxes from the unused properties, which creates a burden for the community. Until redeveloped, brownfields do not provide good tax base (revenues); often, such properties have tax liens. Thus, brownfield development improves local economy.

In the 1990s, state and local governments implemented new brownfield laws that helped businesses to investigate, clean up, and purchase brownfields. Today several cities are cleaning up brownfields.

In 2002 brownfield cleanups in Massachusetts led to the redevelopment of a new hotel in Worcester, a new riverfront industrial park in New Bedford, and a new residential housing facility for the elderly in Somerville. Three other communities in Massachusetts are developing office, laboratory, and manufacturing space, 200 new housing units, and parks from brownfield sites.

Since 1996 Portland, Oregon, has been working on plans to foster the restoration and reuse of contaminated land and to promote the revitalization of neighborhoods within the city. Public and private partnerships within Portland have cleaned up and recycled hundreds

Before: A former contaminated brownfield is being cleaned up by developers to make way for a marina, shops, and offices. (Courtesy of the Golf Foundation of Rhode Island)

After: A cleaned up brownfield was redeveloped into a marina with offices and shops. (Courtesy of the Golf Foundation of Rhode Island)

of hectares of contaminated property. One property, an old battery manufacturing plant that had been abandoned for many years, now is a work center for job training for individuals with developmental disabilities. The cleanups have helped the local economy by creating thousands of jobs. Due to the success and popularity of the brownfield program, in 2002 the U.S. government allocated more funds for cleanup operations.

Other Strategies for Sustainable Communities

There are strategies other than developing greenbelts, purchasing land, and cleaning up brownfields to curtail sprawl and develop more sustainable communities: clustered development and cohousing are two of them.

Swedish Housing on a Brownfield

In Malmö, Sweden, a 27-unit apartment complex was built on a brownfield, formerly the site of an auto manufacturer. The apartments overlook the Danish coast. Before construction began, the city treated and removed soil from the polluted site and filled it in with about two meters (6 feet) of clean soil. The grounds were landscaped and replanted with various plant species which remove heavy metals from the soil, a process known as phytoremediation.

Phytoremediation plants extract some water or soil pollutants still left in the soil. One plant called riverbirch is used to draw toxic metals into its trunk. Grasses are used to trap toxic compounds into their roots. The photoremediation plants continue to decontaminate the soil around the complex. Periodically, the plants are removed from the site and destroyed.

The Swedish housing development also uses solar collectors to convert energy to heat, and nearby wind turbines are used to supply the apartments with electricity. Grasses and other plants are planted on the walls and roofs to provide overhead insulation and to slow down flooding conditions during heavy storms. The apartment building also recycles its own water into a rebuilt marsh nearby.

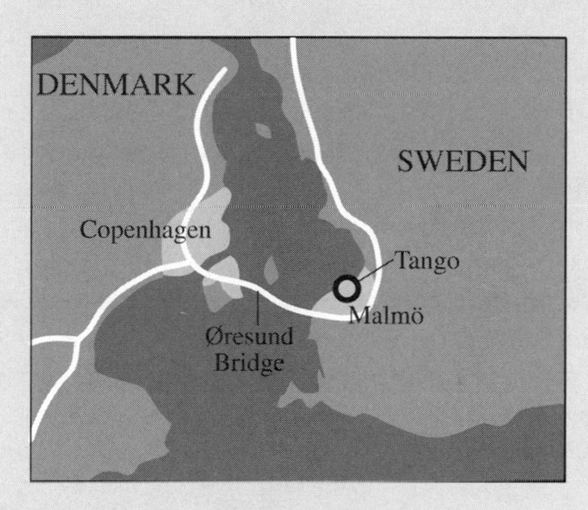

Figure 6-1 • In Malmö, Sweden, the Tango apartment complex uses a variety of alternative energy sources that include wind turbines and photovoltaic systems. The apartments were built on a brownfield, formerly the site of an auto manufacturer.

Clustered Development

The most common type of building in a housing development is usually a large, detached single home located on an expansive lot. This kind of development, leads to sprawl-like conditions. Everything is spread out. One alternative plan to this typical suburban development is to build dwellings clustered together.

Clustered development includes building attached homes which allow more people per hectare (acre) of land than building detached single homes. The clustered community is integrated with stores, schools, professional offices, and businesses in a compact settlement. Clustered development allows more open space, reduces losses of farmland and forests, and cuts down on pollution. Clustered development also has some cost advantages over the spread-out, single, detached home development. The savings of clustered development include fewer costs to

- Construct roads and utility lines (electricity, cable, telephone, gas, and water, etc.)
- Operate and maintain school buses
- Collect refuse
- Run public transportation.

Cohousing

Cohousing, another kind of cluster development, consists of houses or apartments clustered around a common outdoor area. In the common area, there is usually a community house open to all the residents. The community house includes a kitchen and dining or meeting area where the residents share group meals and take care of the business of running the community. Other features in the community house might include a hot tub, exercise room, children's playroom, office, library, craft room, storage room, and guestrooms for visitors.

Cohousing occurs in both rural and city settings. Each person or family owns their own house or unit and shares the responsibility of administering and caring for their community. Sharing is central in cohousing. Shared property includes lawn mowers, tools, laundry facilities, and sometimes automobiles. Community members share outdoor spaces, including gardens and children's play areas, as well as workshops, garages, and a common house.

Cohousing emphasizes resource efficiency, environmental preservation, and a deep sense of community. The layout and construction of the homes emphasize energy efficiency and favor pedestrians over automobiles.

Clustered developments and cohousing projects are found in several areas in the United States. The Hundredfold Farm in Gettysburg, Pennsylvania, is a cohousing community which utilizes cluster development to preserve existing farmland and a tree farm. The farm will be used to grow community food and operate an organic garden. The Marsh Commons in Arcata, California, a cohousing community, employs environmentally appropriate technology to reduce wastes and dependency on cars. The development is built on a reused industrial site adjoining a wildlife refuge situated on a marsh.

One goal of sustainable development strategies is to encourage home ownership for low-income families. The Dade County Jordan Commons program in Florida is designed to provide affordable, quality housing for low-income residents with energy- and water-efficient technologies.

In 1992, when Hurricane Andrew hit Dade County, it proved to be the costliest, most destructive hurricane disaster in U.S. history. Damage costs were estimated to be in excess of $25 billion. Thousands of homes were damaged, and 250,000 people were left homeless. The Dade County Jordan Commons program is alleviating some of the housing shortage by developing a $17 million, 200-home, model community for low-income families left homeless by Hurricane Andrew. The program will feature energy-efficient technologies and incorporate recycling, composting, landscaping, and water-conservation methods. Educational programs will reinforce the use of resource-saving technologies and build community awareness of the importance of conservation.

ENVIRONMENTAL ACTIVISM IN THE COMMUNITY

Turn the Tide

Local communities are taking action to conserve energy and protect the environment. In the Washington, D.C., area, many families and communities belong to a program called Turn the Tide. The program recommends nine relatively simple actions which can be performed by any individual or family to conserve water and energy, reduce the emission of gases that might contribute to global warming, and help save wildlife and forest habitats. Action 1, called "Skip a car trip each week," is a good step in reducing local air pollution and congestion problems. The program recommends that Turn the Tide participants reduce their driving by 32 kilometers (20 miles) a week. A Turn the Tide online calculator computes carbon dioxide emission savings based on the mileage reduction and the average miles per gallon of the car used. Other actions include eating less beef, eliminating lawn and garden pesticides, and replacing standard lightbulbs with energy-saving, compact, fluorescent bulbs. The group has about 1,000 participating members.

EcoTeams

In Philadelphia, by 2002, a group called EcoTeams had diverted an estimated 30 tons of waste from landfills since 1999. EcoTeams, of about five to eight households, meet periodically to develop plans to start a carpool, reduce garbage, conserve water, buy food in bulk, and even stop junk mail delivery. EcoTeams are relatively new to the United States; the first EcoTeams were formed in Europe in 1992.

Local Community Currencies

Some communities, such as Ithaca, New York, and Burlington, Vermont, issue their own currencies to boost their local economy and promote connections within the community. Instead of paying with

DID YOU KNOW?

Local currencies were popular after the Great Depression in the late 1930s when it was hard for people to get cash.

federal dollars, people can trade local currency dollars called scripts for goods such as bread and lunches and services such as grass cutting and baby-sitting. Each script is worth the equivalent of the average hourly wage of the community. The scripts are issued in bills worth $1, $5, and $10. The price of the goods or services depends on the amount of labor involved in producing the goods or services. Nearly 30 communities throughout the United States use local currencies.

Instead of paying with Federal dollars, people in Ithaca, New York, can trade local currency called Ithaca scripts for goods. (© Courtesy of Paul Glover, Ithaca Hours, www.ithacahours.com)

Frederick Law Olmsted (1822–1903)

Central Park, in New York City, was the first large public park or greenbelt developed in the United States. Frederick Law Olmsted was the American architect who designed and constructed Central Park. The park has long, sweeping vistas and scenery; even an outcropping of rocks was left in place to remind visitors of the original landscape. Central Park was designed as a place of peace—a refuge from the stress of the city.

Olmsted also incorporated detailed plans for integrating traffic flow in his landscape designs. He applied his knowledge of topography, geology, botany, and hydrology into his planning of parks. His urban parks, including Central Park, integrated the city and landscape by taking the radical step of routing traffic through the park, not away from it. He did this without disrupting the landscape by physically separating roads either by sinking them out of sight or hiding them behind landforms. In Central Park he included intersections that are sunk about 2.5 meters (eight feet) below the level of the park to minimize the visual and noise pollution of the cars. He created a loop around the park for strolling, running, and parading.

Although well-known throughout the world for his landscaping of New York City's Central Park, Olmsted also designed the landscapes for the U.S. Capitol grounds, the White House, and the Jefferson Memorial. His other works include using natural landscapes for West Point Military Academy, Great Smoky Mountains National Park, and Niagara Falls Reservation. All of his landscape work incorporates the use of native species rather than nonnative ones.

In 1963 Olmsted's home and office in Brookline, Massachusetts, became a National Historic Site. The Olmsted archives contain the most widely researched museum collections in the National Park Service. The archives contain records for nearly 5,000 projects, including national, state, and city parks; arboretums; and school and college campuses. Olmsted's plans include entire park systems for the cities of Seattle, Chicago, Baltimore, and Buffalo.

STRATEGIES FOR A SUSTAINABLE TRANSPORTATION SYSTEM

Traffic Congestion and Air Pollution

Traffic congestion is another major problem of sprawl. In 2002 the Texas Transportation Institute reported that the average person spends 36 hours a year sitting in traffic; in 2000 the national average was 34 hours annually. To get an idea of how bad congestion has become, in 1982, that same person spent only 11 hours in traffic annually. The most congested highways in the country are found in Los Angeles, where residents in 1999 averaged 56 hours a year in traffic.

The Texas Transportation Institute gathered traffic congestion data from 68 urban sites. According to the report, traffic congestion costs an estimated $168 billion annually in lost productivity, and it is expected to triple in coming years, wasting more productivity and fuel and worsening air quality. Billions of dollars a year are spent wasting time and burning gasoline, according to the institute. Other costs involved in congestion and auto travel include traffic deaths and injuries.

In the United States, auto travel accounts for 90 percent of all motorized passenger travel; in Europe, auto travel accounts for 78 percent of all passenger travel. As sprawl increases, the reliance on automobiles and driving will continue to increase also, causing more air pollution and global warming.

Some communities have found a promising new course for handling growth and their transportation problems. Planners refer to these ideas as "livable" or "sustainable" communities. These plans focus on people, rather than on cars. One plan to ease traffic congestion and reduce pollution is to use *mass transit*. In this more efficient form of

DID YOU KNOW?

It is estimated that 67 percent of the land in most cities is devoted to moving, parking, or servicing cars.

TABLE 6-2 The Nation's 10 Worst Urban Areas in Traffic Congestion, 2002

1. Los Angeles
2. San Francisco-Oakland
3. Chicago
4. Washington, D.C.
5. Boston
6. Miami-Hialeah
7. Seattle-Everett
8. Denver
9. San Jose
10. New York-Northeastern N.J.

Source: Texas Transportation Institute.

travel, fewer people use their cars, saving fuel and easing traffic congestion. Mass transit includes buses, commuter rails, light rail trains, van pools, and even bicycles. In the United States, bus travel makes up about 62 percent of all mass transit activities; travel by Amtrak trains accounts for 27 percent; and the rest of passenger travel is done by commuter rails and van pools.

Bus Transit

The bus system with the highest use of any public transit system in the world is located in Curitiba, a large city located on the eastern coast in the southern part of Brazil. Today Curitiba has an urban population of little more than 2 million.

In the early 1970s, Curitiba did not have enough money to build a massive transportation network. It was decided that a bus system would be the least expensive option. The city planners designed a new system in conjunction with road and zoning plans so that the city would not develop sprawl. They established a set of express bus lines radiating from the city center like the spokes of a wheel. The highways had two lanes, designated for buses only. Double- and triple-length buses increased ridership. The city also purchased land along the bus corridors for the development of low-income housing, giving its residents easy access to bus transportation.

People pay a reasonable, flat rate to get anywhere in the city. Over 75 percent of the commuters and shoppers in Curitiba use it, and many of these people report they previously commuted by car. Compared to other South American cities of its size (over 2 million), Curitiba has very low levels of air pollution.

Cleaner Fuels for Buses

Bus transit systems are found in most cities in the world, but many buses run on diesel fuel. Diesel fuel is a petroleum product that is less refined than gasoline. Conventional diesel engines sometimes produce a thick, black exhaust smoke containing *particulates*, which consist primarily of unburned elements of the diesel fuel. According to the EPA, these particulates are harmful to human health. Many cities are beginning to make their buses more sustainable.

Chattanooga, Tennessee, is well known among city planners for its battery-powered electric bus development and as a world leader in electric mass transit. Chattanooga's downtown shuttle integrates electric buses manufactured in Chattanooga. The bus transit system reduces vehicle traffic and improves air quality by shifting riders from their cars to the buses.

Since October 1992, Peoria, Illinois, has been operating the world's largest bus fleet fueled by *ethanol*. The district has set up a new maintenance facility for the ethanol buses and has trained mechanics in the appropriate maintenance procedures, which are slightly different

Chattanooga's downtown bus shuttle service integrates electric buses, which are serviced and manufactured in Chattanooga. (Provided by the Chattanooga Area Chamber of Commerce)

Propane buses are used in Zion National Park. Hydrocarbon propane, a colorless gas easily separated from petroleum and natural gas, is a common fuel. It can be stored in small and large containers. (Courtesy of Hollis Burkhart)

from those for diesel buses. So far, the Greater Peoria Mass Transit District has accumulated data on 842,000 kilometers (522,000 miles) of bus operation on all routes in the city.

Since May 1992, the Metro Dade Transit Agency has run five buses on *methanol*. These bus engines, which run on 100 percent methanol, were among the first methanol engines to become commercially available. The transit agency, which serves the greater Miami, Florida, area, is collecting data on methanol buses as part of a comprehensive experiment with clean-bus technologies. Metro Dade has now accumulated data on more than 160,000 kilometers (100,000 miles) of operation of the methanol engines.

Since 1991, the Metropolitan Transit Authority of Houston (Metro) has run a pilot program of 14 buses on *liquefied natural gas* (LNG). These first-of-a-kind units operate on a mixture of LNG and diesel. Washington, D.C., uses 160 natural gas–powered buses in its bus fleet. These buses emit 90 percent less soot than diesel buses. The city plans to upgrade another 1,200 buses to natural gas.

Refer to Chapter 2 for more information on alternative fuels.

Using the Rails

Rail transit has been keeping sprawl from becoming a major problem in some urban areas. Rail transit includes light rail transit (LRT), mass rapid transit (MRT), and commuter rail systems. LRT, consisting of from one to four electric car trains, is the basic transportation in many medium-sized cities. LRT systems have been built in such cities as Cologne, Germany; Calgary, Canada; Manila, Philippines; San Diego, California; and Portland, Oregon. The Portland LRT system claims to carry the equivalent of two lanes of traffic coming into the downtown area of the city. The public transportation system has made a major contribution to cleaner air by reducing car-dependent commuting. In fact, the downtown air quality in the city improved after the LRT system was built.

Commuter trains, provided by railroad companies, offer regular service from suburban area stations into central cities. The service is ideal for passengers who live in the suburbs and want to work in the city. For example, commuter trains provide regular service for thousands of passengers into New York City each day from Pennsylvania, New Jersey, and Connecticut, as well as from upstate New York and Long Island. Massachusetts, Rhode Island, and New Jersey make use of commuter trains, as well.

The MRT system—also known as rapid transit or metro—a rail system which operates with from 4 to 10 cars, is known for high-performance speed, reliability, and capacity. MRT systems have been built in Toronto, Canada; Stockholm, Sweden; Washington, D.C.; Atlanta, Georgia; and San Francisco, California.

Many large cities use a combination of different rail systems. Singapore is a city-state with a total land area of 650 square kilometers (250 square miles) and a high-population density of 4.1 million. The two train networks in Singapore include the MRT lines and the LRT system. The MRT links the main, large population centers located in the north, south, east, west areas of the city. The LRT provides local transport needs for residents. By integrating the two networks, travel by train is a convenient mode of transport for commuters. Currently about 63 percent of all motorized trips in Singapore are made by public transport every day—3 million trips by buses, 1 million trips by trains, and 1 million trips by taxis.

In Shanghai, China, it takes local residents almost three hours to commute across some parts of the city. The city of 16 million inhabitants is now planning a more effective public transport system which will enable a local resident to commute across the city center in about one hour in 2006. To accomplish this goal, city planners are planning to double the size of the railroad system by building an additional 200 kilometers (120 miles) of track. When the plan is completed, it is expected rail ridership in the city will increase from the current 5 percent rate to 40 percent.

In Japan and France, high-speed trains provide access between major cities within their countries. In Japan, the high-speed train known as the Bullet train covers thousands of kilometers of tracks in excess of 250 kilometers (150 miles) per hour and carries more than 300,000 passengers each day.

Bike Commuting

There are other ways to reduce car-dependent commuting. Millions of people in Tokyo, Japan, ride bicycles to work or to mass transit stations. From there, they take subways or bullet trains to their destination. In other Asian countries, bicycles account for about 50 percent of all personal travel. In the Netherlands, 28 percent of all passenger transport trips are made by bicycling. In many of the towns in the country, the roads are divided into three distinct sections for cars, pedestrians, and bike traffic, the widest lanes are reserved for bicycles.

Bike commuting in the United States is a different story. Although Americans are large purchasers of bicycles, less than 1 percent of them use bicycles to ride to work. Some U.S. cities are now encouraging bike travel for commuters. Davis, California, is probably the leading U.S. city in bicycle commuting. About 30 percent of commuters in Davis ride bicycles to work. There are more than 65 kilometers (40 miles) of bike paths to encourage bike riding.

Palo Alto, California, is another city in which many people commute by bicycle. The city's comprehensive plan includes bicycle boulevards, streets on which bicycles have precedence over cars. In addition, ordinances require private developers to provide amenities for bicyclists. For example, developers must provide bicycle parking, generally at a ratio of 1 bicycle space for every 10 vehicle spaces.

DID YOU KNOW?

In Beijing, China, bicycles outnumber automobiles—not so in many U.S. cities.

Palo Alto promotes bicycling in a variety of other ways. City employees using private bicycles for city business are reimbursed a certain amount of money per mile traveled, and they can use city-owned bicycles for commuting. By registering in the city's bicycle/walker program, residents can receive a discount coupon for the purchase of a bicycle and a brochure and maps for local bicycling. They also receive a monthly coupon redeemable at selected locations for merchandise useful in biking or walking.

Worldwide, millions of people use bike paths to commute and for pleasure. (Courtesy of Hollis Burkhart)

Telecommuting

Telecommuting is another way to cut down the commuting workforce by keeping workers at home. More than 2 million Americans are already telecommuting—working away from the office—via telephone, fax, or modem, according to the U.S. Department of Transportation. San Diego, California, began its telecommuting program with 33 employees in 1991 and found that they telecommuted an average of 1.16 days per week, for an annual savings of 16,090 kilometers (10,000 miles) of commuting, 1,892 liters (500 gallons) of fuel, and 182 kilograms (400 pounds) of atmospheric pollutants.

TRAFFIC CONGESTION AND SAFETY

Other issues of a car-dependent society, besides traffic congestion and air pollution, are vehicle accidents, which continue to increase. In 1999 automobile fatalities reached almost 21,000 in the United States. To slow down speeding cars and reduce accidents, traffic planners are working on different strategies and programs.

Traffic Calming

One new traffic safety program, called traffic calming, has been initiated in some major cities to cut down on speeding cars. Traffic calming was first used in Holland to make streets safer for children and pedestrians. The plan requires traffic planners to narrow the streets and roads leading into downtown centers. The narrow streets force drivers to slow down. In some plans, lanes of traffic are removed and replaced with wider sidewalks for more pedestrian travel. Speed bumps are also installed to slow down speeding vehicles. Crosswalks are extended out from the sidewalks, which shortens the distance pedestrians need to cross the street. Where implemented, the benefits of traffic calming have decreased accidents.

Traffic Circles

Some traffic planners have replaced dangerous intersections with roundabouts. The roundabout, which looks like a modified traffic circle or rotary, redistributes the volume of traffic. It is designed to keep traffic moving safely, trim congestion, and reduce accidents. Drivers inside the rotary are forced to slow down to speeds of between 25 and

DID YOU KNOW?

ZipCar, a car rental service in Boston, Massachusetts, rents cars on an hourly basis for those who drive infrequently. Fuel, maintenance, and off-street parking are part of the service. Environmental benefits include reducing the number of cars on the roads.

Before: The traffic calming plan in Lake Worth, Florida required traffic planners to narrow the three-lane traffic pattern leading into downtown centers. (By permission of Museum of the City of Lake Worth)

After: The now two-lane traffic pattern forces drivers to slow down. The third lane of traffic was replaced with wider sidewalks or parking spaces. (By permission of Lake Worth Public Library)

30 kilometers (15 and 20 miles) per hour. Other cars and vehicles approaching the roundabout must also slow down and yield to vehicles already in the rotary. The vehicles in the rotary keep circulating and moving safely along while vehicles at the entrance section of the roundabout wait for their turn to enter. In 2002 more than 600 roundabouts had been installed in the United States. The first one was built in Montpelier, Vermont.

Vocabulary

Brownfield Industrial property which is perceived to be undesirable to own or use because of the environmental contamination that exists within the boundaries of the property.

Conservation easements A right that one may have to another's land that is to be conserved only and not developed.

Ethanol Alcohol produced by the fermentation of sugars.

Greenbelt Area of agricultural land, parkland, or woodland that surrounds a city or town.

Infrastructure The basic facilities, such as roads, bridges, and schools, on which a city or state depends.

Liquified natural gas A gas, either propane or butane, produced by refining crude petroleum.

Mass transit Transportation system that serves a large number of people.

Methanol Alcohol manufactured by coal, waste wood, or natural gas.

Particulates Particles of matter emitted into the air by motor vehicles and industrial processes.

Urban Characteristic of a city.

Activities for Students

1. Consider the reasons for urban sprawl. Why would people living in rural areas be attracted to cities? Should governments set up programs to encourage people to live in rural areas?
2. Create a diagram for a clustered community. What would be an ideal way, mindful of environmental constraints, to share space in a residential setting?
3. Think about the Turn the Tide program in Washington, D.C. Create a list of nine ways to help your family to live in an environmentally sustainable manner, and post the list in your home.
4. In one week, track how much time you spend in a car. Investigate public transportation in your area and consider alternative ways to get to the places that you travel to each day.

Books and Other Reading Materials

Benfield, F. Kaid, Donald D. Chen, and Matthew D. Raimi. *Once There Were Greenfields: How Urban Sprawl Is Undermining America's Environment, Economy, and Social Fabric.* New York: Natural Resource Defense Council, 1999.

Benfield, F. Kaid, Jutka Terris, and Nancy Vorsanger. *Solving Sprawl: Models of Smart Growth in Communities across America.* New York: Natural Resource Defense Council, 2001.

Beveridge, C. E., et al., eds. *Frederick Law Olmsted: Designing the American Landscape.* New York: Rizzoli Books, 1995.

Davis, Bertha. *Poverty in America: What We Do about It.* New York: Franklin Watts, 1999.

Duany, Andres, Elizabeth Plater-Zyberk, and Jeff Speck. *Suburban Nation: The Rise of Sprawl and the Decline of the American Dream*. New York: North Point Press, 2001.

Rome, Adam. *The Bulldozer in the Countryside: Suburban Sprawl and the Rise of American Environmentalism*. Cambridge, England: Cambridge University Press, 2001.

Websites

Frederick Law Olmsted, http://fredericklawolmsted.com

NASA, http://science.nasa.gov/headlines/y2002/11oct_sprawl.htm

National Geographic: Earth Pulse, http://www.nationalgeographic.com/earthpulse/

Natural Resources Defense Council, http://www.nrdc.org/

Sustainable communities network, http://www.sustainable.org/

U.S. Department of Energy: Smart Communities Network, http://www.sustainable.doe.gov/buildings/gbedtoc.shtml; http://www.sustainable.doe.gov/greendev/artpub.shtml

U.S. Environmental Protection Agency, http://www.epa.gov/epaoswer/non-hw/reduce/catbook/model.htm

U.S. Environmental Protection Agency: Brownfields, Cleanup, and Redevelopment, http://www.epa.gov/brownfields/

U.S. Environmental Protection Agency: Household hazards, http://www.epa.gov/epaoswer/nonhw/muncpl/reduce.htm#hhw

CHAPTER 7

A Sustainable Society Requires Activism

As noted in the previous chapters, there are businesses, government agencies, consumer groups, community activists, nonprofit organizations, and environment groups working on plans and programs for creating a sustainable society. These groups are reducing consumption, writing and enforcing environmental laws, recycling and reusing waste products, conserving energy, preserving wildlife habitats, and exploring new ways to develop a more ecological economy.

The job of creating a sustainable society, however, should not be a task just for large organizations and governments. Individuals must also take full responsibility for the economic, social, and environmental consequences of their actions. Too often, men and women have a self-centered philosophy giving little thought to the effects of their consumption and existence on the natural resources of their environment. They seem to believe that Earth has unlimited natural resources, wastes are tolerable and expected, and overconsumption helps the economy. Another human-centered belief is that life's value and progress are measured solely by one's own material goods and wealth; furthermore, technology can solve all of our environmental and social problems.

Hazel Wolf (1898–2000) was an environmental activist and a longtime crusader for the Seattle Chapter of the National Audubon Society. In honor of her one-hundredth birthday, the Audubon Society (www.seattleaudubon.org) created a Hazel Wolf "Kids for the Environment" fund to promote environmental appreciation for young people. (Courtesy of Greg Kinney)

The greatest obstacle to achieving a sustainable future in a democratic society is overcoming a feeling of helplessness. We believe the problems are too great for us, as individuals, to solve. As a result, there is a sense of apathy, and we leave all the decision-making processes to others. Can one person make a difference? The answer is yes—if we accept individual responsibility, if we want to be active in community affairs, and if we make sustainable lifestyle choices.

Today, there are individuals who, on their own, have identified environmental and social issues in their neighborhoods and communities and have developed the courage to do something about these issues. These people, who are called *environmental activists*, are devoted Earth

Cathrine Sneed is the founder and director of The Garden Project (www.gardenproject.org) in San Francisco. The mission of The Garden Project is to provide structure and support to former prisoners through on the job training in gardening and tree care, counseling, and assistance in continuing education. The United States Department of Agriculture called The Garden Project "one of the most innovative and successful community-based crime prevention programs in the country." Since The Garden Project began in 1992, more than 3,500 former offenders have gone through the program; 75 percent of The Garden Project apprentices do not return to jail. (Courtesy of Cathrine Sneed)

stewards who respect the natural world and the welfare of humans and other species.

The many thousands of environmental activists include a diverse array of revolutionaries, crusaders, and reformers who devote themselves to issues ranging from global deforestation and chemical pollution to environmental racism and the dumping of toxic wastes in their local communities. This chapter profiles several of these men and women, of all age groups, who have taken action to bring about change—living proof that each one of us can make a difference.

YOUNG ENVIRONMENTAL ACTIVISTS

Environmental activists are not limited to any particular age group. A good number of them are teenagers and even younger. In 1990, when Ocean Robbins was 16 years old, he cofounded with Ryan Eliason, age 18, Youth for Environmental Sanity (YES!). The goals of YES! are to educate, inspire, and empower young people to take positive action for healthy people and a healthy planet. Since 1990 Robbins served as director for five years and is now the organization's president. The program has reached 600,000 students in 1,200 schools in 43 states

When Ocean Robbins was sixteen years old, he cofounded Youth for Environmental Sanity (YES! [www.yesworld.org]). Since then, more than 600,000 thousand students have participated in this program. YES! alumni have organized gang truces, started recycling programs, broken through stereotypes and prejudice barriers in schools, initiated community gardens and park clean-ups, and set up marches and rallies to speak out on issues like environmental justice and the prison industry. (Courtesy of Richard Curtis)

through full school assemblies. YES! has also organized and facilitated 54 week-long summer camps in 6 countries, published 7 youth action guides, and led 150 day-long youth training workshops. The YES! Action Camp focus is on community organization, communication skills, healing racism and sexism, conflict resolution, cultivation of peace in everyone and in the world, press coverage, fund-raising, and public speaking. Robbins has personally organized summer camps and workshops in Singapore, Costa Rica, Russia, Finland, Canada, and across the United States.

Severn Cullis-Suzuki was 12 years old when she and three of her schoolmates raised enough money to attend the United Nations Earth Summit meeting held in Rio de Janeiro, Brazil, in 1992. The Earth Summit was attended by the heads of state of 108 nations, who worked together for the development and adoption of several major agreements designed to benefit the global environment. Among the major agreements of the summit was a document known as Agenda 21, which provides a plan of action for achieving sustainable development. Cullis-Suzuki spoke for six minutes to the delegates urging them to work hard to resolve global environmental issues. She received a standing ovation.

Cullis-Suzuki, a native of Vancouver, British Columbia, Canada, has been active in environmental work since kindergarten. At the age of nine, she started the Environmental Children's Organization which organized cleanup projects at the local beaches. She has worked with native peoples in British Columbia, Southeast Asia, and the Amazon to protect threatened forests from logging. Cullis-Suzuki is a

Severn Cullis-Suzuki, a native of Vancouver, British Columbia, has been active in environmental work since she was in kindergarten. She received the United Nations Environmental Global 500 Award in 1993. (© Jeff Topham)

regular speaker at schools, corporations, conferences, and international gatherings on the necessity of changing our values, listening to children, and behaving as if their future matters. As a television host and presenter, she has participated in a number of programs in Canada, the United States, and Britain. She has written many articles on environmental issues and has published a book. She received the United Nations Environmental Program Global 500 Award in 1993.

In the late 1990s, three teenagers in South Central Los Angeles, California, caught the attention of the media for their community work as environmental activists. Nevada Dove, Fabiola Tostado, and Maria Perez appeared on CNN television and in newspapers and magazines including *Time* and *Seventeen*. The media reported the hard work of the three teenagers in campaigning against the opening of a neighborhood school built across the street from a Superfund site containing chromium, a toxic material that affects human health. The students are proud of the fact that they brought positive changes to their community, and all three agree that you can make a difference no matter how young you are or what color your skin is.

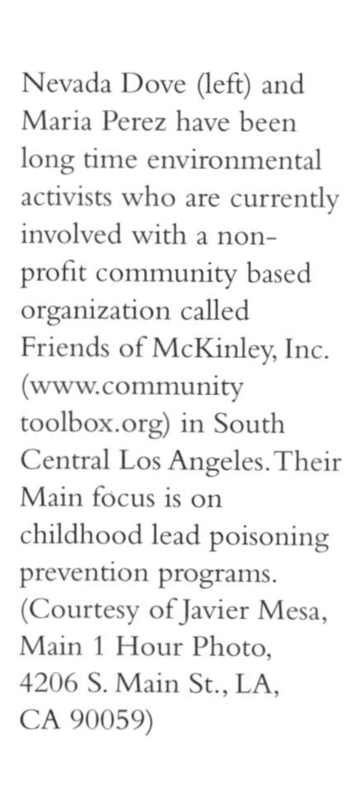

Nevada Dove (left) and Maria Perez have been long time environmental activists who are currently involved with a non-profit community based organization called Friends of McKinley, Inc. (www.communitytoolbox.org) in South Central Los Angeles. Their Main focus is on childhood lead poisoning prevention programs. (Courtesy of Javier Mesa, Main 1 Hour Photo, 4206 S. Main St., LA, CA 90059)

There are other young people who are defending the environment. Simon Jackson, at the age of 17, organized a letter-writing campaign to save the rare, white kermode or spirit bear which lives on an island in British Columbia, Canada. There are only about 100 bears left on the island, and unfortunately it is the only habitat the bears have left. Jackson asked 700 schoolchildren to write to the government asking them to prevent logging companies from working in the bears' natural habitat. Although the bears are still an endangered species, Jackson and his group of letter writers have slowed down the destruction of their habitat.

Chou Yu-sheng lives in Taipei, the capital of Taiwan, a country of about 21 million people. At the age of eight, Yu-sheng organized a group of children in his neighborhood to recycle bottles, cans, and paper wastes by placing recycling bins in his neighborhood.

DID YOU KNOW?

Environmental activist and entomolgist Edward O. Wilson, Jr., has discovered about 300 new ant species. He is a strong activist for species preservation.

ACTIVISM, AN AMERICAN PASTIME

Environmental activists include men and women who have been part of the evolution and growth of the environmental justice movement. Their lifework has been devoted to achieving equality for people of color and for the *disfranchised* and to eliminating the unfair environmental burdens that historically have been placed on their communities. Some of these activists include Robert Bullard, Charles Lee, Lois Gibbs, Randy Hayes, Juana Beatriz Gutiérrez and Denis Hayes.

Robert Bullard is the founder of the Environmental Justice Resource Center at Clark University in Atlanta, Georgia. Bullard's environmental justice career began in 1979 when his wife, attorney Linda McKeever Bullard, represented a group of African American home owners in Houston, Texas, who were opposed to a plan that would locate a municipal landfill in the middle of their backyards. Bullard's book, *Dumping in Dixie*, demonstrates that limited housing and residential options, combined with discriminatory facility-siting practices, contributed to the placement of all types of waste sites in African American communities. Such waste sites include garbage dumps, hazardous-waste landfills, incinerators, lead smelter operations, paper mills, chemical plants, and a host of other polluting industries. These industries have generally followed the path of least resistance, which has been to locate waste sites in economically poor and politically powerless African American communities. This has resulted in a prevalence of brownfields in such communities.

Bullard has taken the environmental justice message to numerous radio, television, and news programs. He has given keynote addresses at major international environmental conferences in the United States, the United Kingdom, Australia, Chile, Brazil, Mexico, Canada, and South Africa. Bullard was the lead author of the position paper that created the Washington Office on Environmental Justice, a people-of-color think tank based in the nation's capital.

Robert Bullard is the Director of the Environmental Justice Resource Center at Clark University in Atlanta, Georgia (http://www.ejrc.cau.edu). His recent book, *Transportation Racism: New Routes to Equity* describes chronic inequality in public transportation. (Courtesy of Robert Bullard and the Environmental Justice Resource Center, Clark Atlanta University)

Lois Gibbs is an environmental activist and founder and executive director of the Center for Health, Environment, and Justice, a clearinghouse for community groups seeking information and assistance in fighting existing or proposed environmental threats. She is well known for her efforts in exposing a hazardous toxic waste dump in Love Canal, New York, in the late 1970s. About 20,000 tons of buried chemicals polluted the ground and the water in a residential area of 800 families. Her grassroots efforts have influenced the national public awareness of toxic wastes in their environment, and her actions have led to the formation of the federal *Superfund* program. Superfund is a common name for the Comprehensive Environmental Response, Compensation, and Liability Act (CERCLA) of 1980, a federal law whose mission is to clean up the worst hazardous and toxic waste site areas on land and water which constitute threats to human health and the environment.

Randy Hayes is the founder and president of the Rainforest Action Network (RAN) situated in San Francisco, California. In little over five years, he has built one of the better known environmental organizations. With more than 40,000 members and 150 affiliates, he continues to make governments and multinational corporations take notice of their actions or lack of action. As director of RAN, Hayes is a leader in the effort to halt the destruction of tropical rain forests and fight for the rights of indigenous people. In 1985 he gathered the who's who of rain forest activists from around the world to create a more effective global movement. Hayes leads a national campaign to confront the overconsumption of wood use, calling for a 75 percent reduction of wood and wood-based paper in the United States within a decade.

Randy Hayes is the founder and president of the Rainforest Action Network (RAN [www.ran.org]) in San Francisco, California. His organization includes 40,000 members whose goal is to halt destruction of tropical rainforests and to protect the rights of indigenous people. Since it was founded in 1985, the Rainforest Action Network has been working to protect tropical rainforests and the human rights of those living in and around those forests. (Courtesy of Randy Hayes, Rainforest Action Network)

Charles Lee is an environmental justice activist and advocate. Lee, a Chinese American, was born in Taiwan and came to the United States when he was eight years old. Lee was motivated by an incident that occurred in Warren County, North Carolina, in 1982. The county was planning to build a landfill for *PCB (polychlorinated biphenyl)* wastes. Many groups felt that the proposed site was racially motivated and posed an environmental health problem for the residents. As a result, there were widespread protests including marches. More than 500 demonstrators were arrested. Although the protestors were unsuccessful in blocking the PCB landfill, they brought national attention to waste facility siting inequities and encouraged African American church and civil rights leaders' support for environmental justice. Lee believes the incident served as the turning point that transformed local environmental issues in a poor minority community into a national movement of environmental justice.

Juana Beatriz Gutiérrez is the cofounder and president of Madres del Este de Los Angeles–Santa Isabel (Mothers of East Los Angeles–Santa Isabel), or MELASI. Gutiérrez and several other women formed MELASI to take a pro-active approach to community improvement.

DID YOU KNOW?

Wangari Maathai, the first female professor at the University of Nairobi, Kenya, started the Green Belt Movement in 1977. This environmental activist and her members have planted more than 10 million trees in several African countries.

Juana Beatriz Gutiérrez is the co-founder and president of Madres del Este de Los Angeles–Santa Isabel (Mothers of East Los Angeles–Santa Isabel). Gutiérrez and her organization take a very pro-active approach to community improvement. (Courtesy of Juana Gutiérrez)

MELASI joined with national environmental groups to oppose the proposed the toxic waste incinerator in a neighboring city of Vernon near East Los Angeles. MELASI began to address quality-of-life issues and became a proactive organization committed to working within the environmental movement on issues of education and community economic self-reliance. This proactive approach led MELASI to network with activists from several other parts of the country.

MELASI initiated a water conservation project with the *Mono Lake* Committee in California. In an effort to preserve the natural habitats from which Southern California draws its water resources, this project focused on reducing the level of water consumption by Southern California residents. Economically, consumers' water bills were reduced by ultra-low flush toilets which were distributed free of charge. In addition, 22 community people were employed full-time with benefits through this program. MELASI also converted a former vacant lot into the Eastside Community Garden and Education Center. El Jardín (The Garden) provides local schoolchildren and teachers with a learning facility. Several events are held throughout the year at the Eastside garden to promote culture and environmentalism as well.

Denis Hayes, who has been an environmental lawyer, and professor of engineering, was the coordinator of Earth Day 2000, 31 years after coordinating the first Earth Day in 1969. The first Earth Day was

FIGURE 7-1 • Mono Lake, California In 1941, a system of aqueducts began diverting water from several streams feeding Mono Lake to the city of Los Angeles, located about 650 kilometers away, to help meet the water needs of its residents. Diversion of this water has significantly altered the Mono Lake ecosystem.

Denis Hayes is president of Bullitt Foundation, an environmental organization. Hayes has been an environmental lawyer, professor of engineering, and was the Coordinator of Earth Day 2000, 31 years after coordinating the first Earth Day in 1969.

rooted in a proposal made by Gaylord Nelson, then a U.S. senator, for a nationwide "Environmental Teach-In."

According to Hayes, "On that first Earth Day, a huge cross section of America came together around a common set of values, prompted perhaps by the most basic biological drive—the drive for survival. Twenty million people took part in that first Earth Day, and it created the impetus for the creation of the EPA and the passage of the Clean Air Act, the Clean Water Act, the Endangered Species Act, and most of the rest of the political framework of modern environmental policy." Hayes is now president of Bullitt Foundation, an environmental organization.

Environmental Pioneers, a Few Short Stories

John Muir (1838–1914), a farmer, inventor, naturalist, explorer, writer, and conservationist, was born on April 21, 1838, in Dunbar, Scotland. Muir is perhaps this country's most famous and influential naturalist and *conservationist*. He taught the people of his time and ours the importance of experiencing and protecting our natural heritage. His words have heightened our appreciation of nature. His personal and determined involvement in the great conservation questions of the day remains an inspiration for environmental activists everywhere.

Marjory Stoneman Douglas (1890–1998), an environmentalist, journalist, playwright, activist, feminist, and independent thinker, was born on April 7, 1890, in Minneapolis, Minnesota. Her name is synonymous with the *Florida Everglades* for her grassroots efforts to protect and preserve this watery region—a region many considered to be a large, useless, and worthless swamp. Taking nearly five years to research and write, *The Everglades: River of Grass* went on sale in November 1947 and sold out in a few months. It became a best-selling book. Her interest in environmental issues was only natural, she told *Time* magazine in 1983. "It's women's business to be interested in the environment. It is an extended form of housekeeping." In 1993 President Bill Clinton invited Douglas to the White House and presented her with the Medal of Honor for her environmental work.

Rachel Louise Carson (1907–1964) was a marine biologist, educator, and writer. From an environmental perspective, Carson's most significant writing was her final book, *Silent Spring*, published in 1962. In this work, she details ways in which many chemicals being manufactured by industry and being used in agriculture could be harmful to organisms in the environment. Among the most notable of these was the pesticide *DDT*, which was widely used

Muir Woods is a National Monument located in Mill Valley, California. Giant redwood trees dominate the 300 acre park. (Courtesy of John D. Mongillo)

Environmental Pioneers, a Few Short Stories (*continued*)

at that time throughout the world. Carson proved that the pesticides were killing not only the pests but also the secondary predators, such as birds, who ate the sprayed insects. In 1972 the U.S. Congress enacted legislation banning further use of DDT in this country and began regulating the use of other chemicals that could potentially harm the environment.

Theodore Roosevelt (1858–1919) was a naturalist, writer, soldier, *preservationist*, and conservationist. He was also the governor of New York and the twenty-sixth president of the United States. Born on October 27, 1858, in New York City, Roosevelt as a young boy collected insects, mammals, and birds; at the age of 13 he took taxidermy lessons to became a taxidermist. He could walk through the local woods and identify birds by their songs and Latin names. During his tenure in the White House, from 1901 to 1909, he designated 150 national forests, the first 51 federal bird reservations, 5 national parks, the first 18 national monuments, the first 4 national game preserves, and the first 21 reclamation projects.

SCHOOLS USING SUSTAINABLE PRACTICES

More than 80,000 primary and secondary schools use environmental and sustainable practices. One school in New York, High School for Environmental Studies, has a recycling center, roof garden, greenhouse, composting center, weather station, and a computerized research library. A public school in Utah offers educational programs on the environment, culture, and sustainability in the Southwest. They emphasize the need to preserve the natural and cultural resources in the Southwest and around the world.

Public schools in Aurora, Colorado, improved energy efficiency in their school buildings. They replaced old weather stripping and caulking and put active solar energy systems into several of the new elementary schools. A Portland, Oregon, high school adapted a program that saves the school thousands of dollars each year and reduces yearly garbage disposal by several tons. The school replaced paper food trays with wax tissue sheets, which use less paper than the trays; recharged used computer cartridges; and used both sides of copy paper for print runs. The Rhode Island Schools Recycling Club consists of many Rhode Island public and private schools. The goal of each school is to separate loose paper from the trash and find an efficient method to transport it to a recycling facility. Recycling saves trees, saves energy, saves water, and reduces garbage loads.

Schools in the United States spend more than two billion dollars a year on energy costs. Some of these schools are looking for alternative and renewable energy sources to cut down on their energy costs. As one example, the town of Spirit Lake, Iowa, uses wind turbines to generate electricity in their schools. In 1993 the town installed one wind turbine to power one school. In October 2001, a second wind turbine was

Students at Pattonville High School in Maryland Heights, Missouri, suggested the use of methane from a nearby landfill to heat their school building. The community and developers initiated a plan to pipe in the methane gas. The school is now heated with methane gas. (Courtesy of Pattonville School District, St. Louis, Missouri)

installed to power electricity to the district's high school, middle school, the administration building, a technical building, a bus barn, and the football stadium's lights. Science students at Pattonville High School in Maryland Heights, Missouri, suggested on idea using methane from a nearby landfill to heat their school building. Based on their suggestion and the work of many others, the community adopted a plan to pipe in the land fill gas to the school. The gas, produced from decaying wastes, is delivered to the school by a 1,100-meter (3,600-foot) pipeline from the landfill. About 100 classrooms and gymnasiums are heated by the methane.

Today, there are thousands of stories of individuals and school groups who are cleaning up beaches, rivers, and streams; planting trees; clearing away trash; recycling; and conserving energy. Individuals and school participation can make a difference in creating a sustainable future.

WHAT CAN I DO?

One person can make a difference, even if it is a small change. A good place to start is to make use of the three environmental Rs: reduce, reuse, and recycle. Cut down on the amount of waste you create, keep and reuse products as long as you can, and recycle those products that cannot be reused, whenever possible.

Here are a few simple tips gleaned from some environmental groups.

1. Take your lunch to school in reusable containers instead of throwaway paper and plastic bags.
2. Encourage your school to stop or reduce the use of throwaway dishes, cups, and utensils.

3. Set up waste receptacles in your school for uneaten food that can be composted on school grounds. Encourage each classroom to have a special bin just for recycling paper.
4. Encourage the school to conserve water by installing low-flow showerheads in schools.
5. Donate clean and reusable clothes and old bicycles to special agencies.
6. Turn off the lights in rooms that are not being used during most of the day and night.
7. Contact local utility companies to conduct an energy audit to identify where the energy is being wasted.
8. Use fluorescent lights because they use much less energy than conventional lights.
9. Grow a plant in your room. It will clean the air by absorbing carbon dioxide and releasing oxygen.
10. Volunteer in your community to collect trash from wooded areas and streams.
11. Look for products with less packaging or wrapping and determine which can be conveniently recycled or reused.
12. Save water by turning off the faucet instead of letting the water run while you brush your teeth. Scrub vegetables in a basin of water—not under running water.

Vocabulary

Conservationist Individual who believes in a national policy that stresses a careful and organized management of some natural resources.

DDT Insecticide used to control malaria-carrying mosquitoes and a variety of insect pests, it is now banned in the United States.

Disfranchise To deprive someone of a privilege, right, or power.

Environmental Activist Person who takes positive action to identify and solve environmental issues.

Florida Everglades National park, established in 1947 on the southwestern tip of the Florida peninsula, which covers and includes islands along the Gulf of Mexico, parts of Big Cypress swamp, and a marsh.

Mono Lake Saltwater lake in California which has been adversely affected by the diversion of water from its feeder streams. Diversion of this water has significantly altered the Mono Lake ecosystem.

PCB (polychlorinated biphenyl) Variety of organic compounds originally used as insulation in electrical parts and equipment.

Preservationist Individual who believes in a national policy that stresses the aesthetic aspects of forests and other wildlife habitats and often favors leaving such areas undisturbed.

Superfund Popular name for the Comprehensive Environmental Response, Compensation, and Liability Act of 1980. The Superfund provides federal funds to finance the cleanup of hazardous waste dumpsites.

Activities for Students

1. Find out about local environmental groups and become involved with their activities.
2. Pick one of the activists listed in this chapter and research his or her life and accomplishments. What made this person interested in environmental protection?
3. Establish an environmental group or club at your school. Create a mission statement, start an informational campaign, and plan environmental awareness events for your school community.

Books and Other Reading Materials

Bullard, Robert D. *Dumping in Dixie: Race, Class, and Enrivonmental Quality*. Boulder, Colo.: Westview Press, 1990.

Carson, Rachel. *Silent Spring*. Boston: Houghton Mifflin, 1962.

Cutright, Paul Russell. *Theodore Roosevelt, the Making of a Conservationist*. Chicago: University of Illinois Press, 1985.

Douglas, Marjory Stoneman. *The Everglades: River of Grass*. Sarasota, Fla.: Pineapple Press, 1997.

Earth Works Project Staff. *Fifty Simple Things You Can Do to Save Earth*. New York: Greenleaf Publications, 1990.

Gottlieb, Robert. *Environmentalism Unbound: Exploring New Pathways for Change*. Cambridge, Mass. MIT Press, 2001.

Hayes, Denis. *The Official Earth Day Guide to Repair the Planet*. Washington, D.C.: Island Press, 2000.

Herrtsgaard, Mark. *Earth Odyssey*. New York: Broadway Books, a Division of Random House, 1998.

Hill, Julia Butterfly and Jessica Hurley. *One Makes the Difference*. San Francisco: Harper, 2002.

Jones, Ellis, Ross Haenfler, and Brett Johnson. *The Better World Handbook: From Good Intentions to Everyday Actions*. Gabriola Island, B.C., Canada: New Society Publishers, 2001.

Lerner, Steven. *Eco-Pioneers: Practical Visionaries Solving Today's Environmental Problems*. Cambridge, Mass.: MIT Press, 1997.

MacEachern, Diane. *Save Our Planet: 750 Everyday Ways You Can Help Clean Up the Earth*. New York: Bantam Doubleday Dell Publishing, 1995.

Markle, Sandra. *After the Spill: The Exxon Valdez Disaster, Then and Now*. New York: Walker, 1999.

Milbrath, Lester. *Learning to Think Environmentally: While There Is Still Time*. New York: University of New York Press, 1996.

Mongillo, John, and Bibi Booth, ed. *Environmental Activists*. Westport, Conn.: Greenwood Press, 2001.

Rasmussen, Larry. *Earth Community, Earth Ethics*. Maryknoll, N.Y.: Orbis Books, 1998.

Rees, William, and Phil Testemale. *Our Ecological Footprint: Reducing Human Impact on the Earth*. Gabriola Island, B.C., Canada: New Society Publishers, 1995.

Rifkin, Jeremy. *Green Lifestyle Guide*. New York: Henry Holt, 1990.

Shutkin, William, and David Ross Brower. *The Land That Could Be: Environmentalism and Democracy in the Twenty-First Century*. Cambridge, Mass.: MIT Press, 2001.

Silver, Debbie, and Bernadette Vallely. *Earth Action!: A Guide to Saving the Planet*. New York: Farrar, Straus, & Giroux, 1991.

Websites

Clean Air Challenge, http://www.clair.org

Encyclopedia of Sustainable Development, http://www.doc.mmu.ac.uk/aric/esd/menu.html

Endangered Species Program of the U.S. Fish and Wildlife Service, http://endangered.fws.gov/index.html

Environmental Ethics, http://www.cep.unt.edu/

Environmental Justice Information Page, University of Michigan, http://www-personal.umich.edu/~jrajzer/nre/

Environmental News Service, www.enn.org

Human Rights Watch, www.hrw.org

Is it clear or hazy outside? See it now with live camera shots from locations across the northeast, http://www.hazecam.net

Sierra Club, www.sierraclub.org

Union of Concerned Scientists, www.ucsusa.org

United Nations Development Program, www.undp.org

United Nations Food and Agriculture Organization, www.fao.org

World Bank, www.worldbank.org

World Resources Institute, www.wri.org

World's largest database of animal protection societies, http://worldanimalnet.org/

Worldwatch Institute, www.worldwatch.org

Appendix A: Environmental Timeline, 1620–2004

Environmentalists and activists appear in **boldface**.

1620 to 1860 Erosion becomes a major problem on many American farms. Fields are abandoned. Rivers and streams are filled with silt and mud. The publication of farm journals is initiated by early soil conservationists to improve farming methods.

1748 Jared Eliot, a minister and doctor of Killingsworth, Connecticut, writes the first American book on agriculture to improve crops and to conserve soil.

1824 Solomon and William Drown of Providence, Rhode Island, publish *Farmer's Guide* which discusses erosion and its causes and remedies. A year later, John Lorain, of the Philadelphia Agricultural Society, publishes a book devoted to the prevention of soil erosion in which he discusses methods such as using grass as an erosion-control crop.

1827 John James Audubon begins publication of *Birds of America*.

1830 George Catlin launches his great western painting crusade to document Native American peoples.

1845 Henry David Thoreau moves to Walden Pond to observe the fauna and flora of Concord, Massachusetts.

1847 U.S. Congressman **George Perkins Marsh** of Vermont delivers a speech calling attention to the destructive impact of human activity on the land.

1849 The U.S. Department of the Interior (DOI) is established.

1857 Frederick Law Olmsted develops the first city park: New York City's Central Park.

1859 British naturalist Charles Darwin publishes *The Origin of the Species by Means of Natural Selection*. In time the theory of evolution presented in the book becomes the most widely accepted theory of evolution.

1866 German biologist Ernst Haeckel introduces the term *ecology*.

1869 John Muir moves to the Yosemite Valley.

Geologist and explorer John Wesley Powell travels the Colorado River through the Grand Canyon.

1872 Yellowstone National Park is established as the first national park of the United States in Yellowstone, Wyoming.

U.S. legislation: Passage of the Mining Law permits individuals to purchase rights to mine public lands.

1876 The Appalachian Mountain Club is founded.

1879 The U.S. Geological Survey (USGS) is formed.

1882 The first hydroelectric plant opens on the Fox River in Wisconsin.

1883 Krakatoa, a small island of Indonesia, is virtually destroyed by a volcanic explosion.

1890 Denmark constructs the first windmill for use in generating electricity.

Sequoia National Park, Yosemite National Park, and General Grant National Park are established in California.

1891 U.S. legislation: Passage of Forest Reserve Act provides the basis for a system of national forests.

1892 John Muir, Robert Underwood Johnson, and William Colby are cofounders of the Sierra Club, in Muir's words, to "do something for wildness and make the mountains glad."

1893 The National Trust is founded in the United Kingdom. The group purchases land deemed of having natural beauty or considered a cultural landmark.

1895 Founding of the American Scenic and Historic Preservation Society.

1898 Cornell University establishes the first college program in forestry.

Gifford Pinchot becomes head of the U.S. Division of Forestry (now the U.S. Forest Service) and serves until 1910. Under President

Theodore Roosevelt, many of Pinchot's ideas became national policy. During his service, the national forests increase from 32 in 1898 to 149 in 1910, a total of 193 million acres.

1899 The River and Harbor Act bans pollution of all navigable waterways. Under the act, the building of any wharves, piers, jetties, and other structures is prohibited without congressional approval.

1900 U.S. legislation: Passage of Lacey Act makes it unlawful to transport illegally killed game animals across state boundaries.

1902 U.S. legislation: Passage of Reclamation Act establishes the Bureau of Reclamation.

1903 First federal U.S. wildlife refuge is established on Pelican Island in Florida.

1905 The National Audubon Society, named for wildlife artist John James Audubon, is founded.

1906 Yosemite Valley is incorporated into Yosemite National Park.

1907 International Association for the Prevention of Smoke is founded. The group's name later changes several times to reflect other concerns over causes of air pollution.

Gifford Pinchot is appointed the first chief of the U.S. Forest Service.

1908 The Grand Canyon is set aside as a national monument.

Chlorination is first used at U.S. water treatment plants.

President Theodore Roosevelt hosts the first Governors' Conference on Conservation.

1914 The last passenger pigeon, Martha, dies in the Cincinnati zoo.

1916 The National Park Service (NPS) is established.

1918 Hunting of migratory bird species is restricted through passage of the Migratory Bird Treaty Act. The act supports treaties between the United States and surrounding nations.

Save-the-Redwoods League is created.

1920 U.S. legislation: Passage of the Mineral Leasing Act regulates mining on federal lands.

1922 The Izaak Walton League is organized under the direction of **Will H. Dilg**.

1924 Environmentalist **Aldo Leopold** wins designation of Gila National Forest, New Mexico, as first extensive wilderness area.

Marjory Stoneman Douglas, of the *Miami Herald*, writes newspaper columns opposing the draining of the Florida Everglades.

Bryce Canyon National Park is established in Utah.

1925 The Geneva Protocol is signed by numerous countries as a means of stopping use of biological weapons.

1928 The Boulder Canyon Project (Hoover Dam) is authorized to provide irrigation, electric power, and a flood-control system for Arizona and Nevada communities.

1930 Chlorofluorocarbons (CFCs) are deemed safe for use in refrigerators and air conditioners.

1931 France builds and makes use of the first Darrieus aerogenerator to produce electricity from wind energy.

Addo Elephant National Park is established in the Eastern Cape region of South Africa to provide a protected habitat for African elephants.

1932 Hugh Bennett is given the opportunity to put his soil conservation ideas into practice to help reduce soil erosion. He becomes the director of the Soil Erosion Service (SES) created by the Department of Interior.

1933 The Tennessee Valley Authority (TVA) is formed.

The Civilian Conservation Corps (CCC) employs more than 2 million Americans in forestry, flood control, soil erosion, and beautification projects.

1934 The greatest drought in U.S. history continues. Portions of Texas, Oklahoma, Arkansas, and several other midwestern states are known as the "Dust Bowl."

U.S. legislation: Passage of Taylor Grazing Act regulates livestock grazing on federal lands.

1935 The Soil Conservation Service (SCS) is established.

The Wilderness Society is founded.

1936 The National Wildlife Federation (NWF) is formed.

1939 David Brower produces his first nature film for the Sierra Club, called *Sky Land Trails of the Kings*. In the same year, Brower, who is an excellent climber, completes his most famous ascent, Shiprock, a volcanic plug which rises 1,400 feet from the floor of the New Mexico desert.

1940 The U.S. Wildlife Service is established to protect fish and wildlife.

U.S. legislation: President Franklin Roosevelt signs the Bald Eagle Protection Act.

1945 The United Nations (UN) establishes the Food and Agriculture Organization (FAO).

1946 The International Whaling Commission (IWC) is formed to research whale populations.

The U.S. Bureau of Land Management (BLM) and the Atomic Energy Commission (AEC) are created.

1947 Marjory Stoneman Douglas publishes *The Everglades: River of Grass* and serves as a member of the committee that gets the Everglades designated a national park.

1948 The UN creates the International Union for the Conservation of Nature (IUCN) as a special environmental agency.

An air pollution incident in Donora, Pennsylvania, kills 20 people; 14,000 become ill.

U.S. legislation: Passage of Federal Water Pollution Control Law.

1949 Aldo Leopold's *A Sand County Almanac* is published posthumously.

1950 Oceanographer **Jacques Cousteau** purchases and transforms a former minesweeper, the *Calypso*, into a research vessel which he uses to increase awareness of the ocean environment.

1951 Tanzania begins its national park system with the establishment of the Serengeti National Park.

1952 Clean air legislation is enacted in Great Britain after air pollution–induced smog brings about the deaths of nearly 4,000 people.

David Brower becomes the first executive director of the Sierra Club.

1953 Radioactive iodine from atomic bomb testing is found in the thyroid glands of children living in Utah.

1955 U.S. legislation: Passage of the Air Pollution Control Act, the first federal legislation designed to control air pollution.

1956 U.S. legislation: Passage of the Water Pollution Control Act authorizes development of water-treatment plants.

1959 The Antarctic Treaty is signed to preserve natural resources of the continent.

1961 The African Wildlife Foundation (AWF) is established as an international organization to protect African wildlife.

1962 Rachel Carson publishes *Silent Spring*, a groundbreaking study of the dangers of DDT and other insecticides.

Hazel Wolf joins the National Audubon Society in Seattle, Washington, and plays a prominent role in local, national, and international environmental efforts during her lifetime.

1963 The Nuclear Test Ban Treaty between the United States and the Soviet Union stops atmospheric testing of nuclear weapons.

U.S. legislation: Passage of the first Clean Air Act (CAA) authorizes money for air pollution control efforts.

1964 Hazel Henderson organizes women in a local play park in New York City and starts a group called Citizens for Clean Air, the first environmental group, she believes, east of the Mississippi. She built Citizens for Clean Air from a very small group to a membership of 40,000. Two years later, 80 people died in New York City from air pollution–related causes during four days of atmospheric inversion.

U.S. legislation: Passage of the Wilderness Act creates the National Wilderness Preservation System.

1965 U.S. legislation: Passage of the Water Quality Act authorizes the federal government to set water standards in absence of state action.

1966 Eighty people in New York City die from air pollution–related causes.

1967 The *Torey Canyon* runs aground spilling 175 tons of crude oil off Cornwall, England.

Dian Fossey establishes the Karisoke Research Center in the Virunga Mountains, within the Parc National des Volcans in Rwanda to study endangered mountain gorillas.

The Environmental Defense Fund (EDF) is formed to lead an effort to save the osprey from DDT.

1968 U.S. legislation: Passage of the Wild and Scenic Rivers Act and the National Trails System Act identify areas of great scenic beauty for preservation and recreation.

Paul Ehrlich publishes *The Population Bomb*.

1969 Wildlife photographer Joy Adamson establishes the Elsa Wild Animal Appeal, an organization dedicated to the preservation and humane treatment of wild and captive animals.

Greenpeace is created.

Blowout of oil well in Santa Barbara, California, releases 2,700 tons of crude oil into the Pacific Ocean.

U.S. legislation: Passage of the National Environmental Policy Act (NEPA) requires all federal agencies to complete an environmental impact statement for any dam, highway, or other large construction project undertaken, regulated, or funded by the federal government.

The Friends of the Earth (FOE) is founded in the United States.

John Todd, **Nancy Jack Todd**, and Bill McLarney are the cofounders of the New Alchemy Institute in Cape Cod, Massachusetts. The institute begins to pioneer a new way of treating sewage and other wastes.

1970 **Denis Hayes** is the national coordinator of the first Earth Day, which is celebrated on April 22.

Construction of the Aswan High Dam on the Nile River in Egypt is completed.

U.S. legislation: Passage of an amended Clean Air Act (CAA) expands air pollution control.

The U.S. Environmental Protection Agency (EPA) is established.

1971 Canadian primatologist Biruté Galdikas begins her studies of orangutans through the Orangutan Research and Conservation Project in Borneo.

The United Nations Educational, Scientific and Cultural Organization (UNESCO) establishes the Man and the Biosphere Program, developing a global network of biosphere reserves.

1972 The Biological and Toxin Weapons Convention is adopted by 140 nations to stop the use of biological weapons.

The EPA phases out the use of DDT in the United States to protect several species of predatory birds. The ban builds on information obtained from Rachel Carson's 1962 book, *Silent Spring*.

U.S. legislation: Passage of the Water Pollution Control Act, the Coastal Zone Management Act (CZMA), and the Environmental Pesticide Control Act.

Oregon passes the first bottle-recycling law.

1973 Norwegian philosopher Arne Naess coins the term *deep ecology* to describe his belief that humans need to recognize natural things for their intrinsic value, rather than just for their value to humans.

The Convention on International Trade in Endangered Species of Wild Fauna and Flora (CITES) is signed by more than 80 nations. The Endangered Species Act of the United States also is enacted.

Congress approves construction of the 1,300-kilometer pipeline from Alaska's North Slope oil field to the Port of Valdez.

An Energy crisis in the United States arises from an Arab oil embargo.

A collision and resulting explosion between the *Corinthos* oil tanker and the *Edgar M. Queeny* releases 272,000 barrels of crude oil and other chemicals into the Delaware River near Marcus Hook, Pennsylvania.

1974 Scientists report their discovery of a hole in the ozone layer above Antarctica.

U.S. legislation: Passage of the Safe Drinking Water Act sets standards to protect the nation's drinking water. The EPA bans most uses for aldrin and dieldrin and disallows the production and importation of these chemicals into the United States.

1975 Unleaded gas goes on sale. New cars are equipped with antipollution catalytic converters.

The EPA bans use of asbestos insulation in new buildings.

Edward Abbey publishes *The Monkey Wrench Gang*, a novel detailing acts of ecotage as a means of protecting the environment.

1976 *Argo Merchant* runs aground releasing 25,000 tons of fuel into the Atlantic Ocean near Nantucket, Rhode Island.

National Academy of Sciences reports that CFC gases from spray cans are damaging the ozone layer.

U.S. legislation: Passage of the Resource Conservation and Recovery Act empowers the EPA to regulate the disposal and treatment of municipal solid and hazardous wastes. The Toxic Substances Control Act and the Resource Conservation and Recovery Act are enacted.

Fire aboard the *Hawaiian Patriot* releases nearly 100,000 tons of crude oil into the Pacific Ocean.

1977 The Green Belt Movement is begun by Kenyan conservationist Wangari Muta Maathai on World Environment Day.

Blowout of Ekofisk oil well releases 27,000 tons of crude oil into the North Sea.

Construction of the Alaska pipeline, the 1,300-kilometer pipeline that carries oil from

Alaska's North Slope oil field to the Port of Valdez, is completed at a cost of more than $8 billion.

U.S. legislation: Passage of the Surface Mining Control and Reclamation Act.

The Department of Energy (DOE) is created.

1978 The *Amoco Cadiz* tanker runs aground spilling 226,000 tons of oil into the ocean near Portsall, Brittany.

People living in the Love Canal community of New York are evacuated from the area to reduce their exposure to chemical wastes which have surfaced from a canal formerly used as a dump site.

Rainfall in Wheeling, West Virginia, is measured at a pH of 2, the most acidic rain yet recorded.

Aerosols with fluorocarbons are banned in the United States.

The EPA bans the use of asbestos in insulation, fireproofing, or decorative materials.

1979 British scientist **James E. Lovelock** publishes *Gaia: A New Look at Life on Earth.*

Collision of the *Atlantic Empress* and the *Aegean Captain* releases 370,000 tons of oil into the Caribbean Sea.

The Convention on Long-Range Transboundary Air Pollution (LRTAP) is signed by several European nations to limit sulfur dioxide emissions which cause acid rain problems in other countries.

The Three Mile Island Nuclear Power Plant in Pennsylvania experiences near-meltdown.

The EPA begins a program to assist states in removing flaking asbestos insulation from pipes and ceilings in school buildings throughout the United States.

The EPA bans the marketing of herbicide Agent Orange in the United States.

1980 Debt-for-nature swap idea is proposed by Thomas E. Lovejoy: nations could convert debt to cash which would then be used to purchase parcels of tropical rain forest to be managed by local conservation groups.

Global Report to the President addresses world trends in population growth, natural resource use, and the environment by the end of the century, and calls for international cooperation in solving problems.

U.S. legislation: Passage of the Comprehensive Environmental Response, Compensation, and Liability Act (Superfund) and the Low Level Radioactive Waste Policy Act.

1981 Earth First!, a radical environmental action group that resorts to ecotage to gain its objectives, formed.

Lois Gibbs founds the Citizens' Clearinghouse for Hazardous Wastes, later named the Center for Health, Environment, and Justice (CHEJ).

1982 U.S. legislation: Passage of the Nuclear Waste Policy Act.

1983 A film of **Randy Hayes**, *The Four Corners, a National Sacrifice Area*, wins the 1983 Student Academy Award for the best documentary. The film documents the tragic effects of uranium and coal mining on Hopi and Navajo Indian lands in the American Southwest.

The residents of Times Beach, Missouri, are ordered to evacuate their community. Investigations of Times Beach in the 1980s disclosed the fact that oil contaminated with dioxin, a highly toxic substance, had been used to treat the town's streets.

Cathrine Sneed founds and acts as director of the Garden Project in San Francisco. The Garden Project, a horticulture class for inmates of the San Francisco County Jail, uses organic gardening as a metaphor for life change. The U.S. Department of Agriculture calls the project "one of the most innovative and successful community-based crime prevention programs in the country."

1984 Toxic gases released from the Union Carbide chemical manufacturing plant in Bhopal, India kill an estimated 3,000 people and injure thousands of others.

The Jane Goodall Institute (JGI) is founded.

The British tanker *Alvenus* spills 0.8 million gallons of oil into the Gulf of Mexico.

U.S. legislation: Passage of the Hazardous and Solid Waste Amendments.

1985 Concerned Citizens of South Central Los Angeles becomes one of the first African American environmental groups in the United States. **Julia Tate** serves as the executive director. The organization's goal is to provide a better quality of life for the residents of this Los Angeles community. **Maria Perez**, **Nevada Dove**, and **Fabiola Tostado** later join the group and are known as the Toxic Crusaders.

Huey D. Johnson becomes the founder and president of the Resource Renewal Institute

(RRI), a nonprofit organization located in California. Johnson suggests that green plans is the path countries should take to respond to environmental decline. Green plans treat the environment as it really exists—a single, interconnected ecosystem that can be safeguarded for future generations only through a systemic, long-range plan of action.

Scientists of the British Antarctica Survey discover the ozone hole. The hole, which appears during the Antarctic spring, is caused by the chlorine from CFCs.

Juana Gutiérrez becomes president and founder of Mothers of East Los Angeles, Santa Isabel Chapter (Madres del Este de Los Angeles—Santa Isabel) (MELASI) whose mission is to fight against toxic dumps and incinerators and also to take a proactive approach to community improvement.

Primatologist Dian Fossey is discovered murdered in her cabin at the Karosoke Research Center she founded. Her death is attributed to poachers.

While protesting nuclear testing being conducted by France in the Pacific Ocean, the *Rainbow Warrior* (a boat owned by Greenpeace) is sunk in a New Zealand harbor by agents of the French government.

U.S. legislation: Passage of the Food Security Act.

1986 Tons of toxic chemicals stored in a warehouse owned by the Sandoz pharmaceutical company are released into the Rhine River near Basel, Switzerland. The effects of the spill are experienced in Switzerland, France, Germany, and Luxembourg.

An explosion destroys a nuclear power plant in Chernobyl, Ukraine, immediately killing more than 30 people and leading to the permanent evacuations of more than 100,000 others.

Bovine spongiform encephalopathy (BSE), a neurodegenerative illness of cattle, also known as mad cow disease, comes to the attention of the scientific community when it appears in cattle in the United Kingdom.

U.S. legislation: Passage of the Emergency Response and Community Right-to-Know Act and the Superfund Amendments and Reauthorization Act (SARA).

1987 The Montreal Protocol, an international treaty that proposes to cut in half the production and use of CFCs, is approved by more than 30 nations.

The world's fourth largest lake, the Aral Sea of Asia, is divided in two as a result of the diversion of water from its feeder streams, the Syr Darya and Amu Darya rivers.

The *Mobro*, a garbage barge from Long Island, New York, travels 9,600 kilometers in search of a place to offload the garbage it carries.

1988 Use of ruminant proteins in the preparation of cattle feed is banned in the United Kingdom to prevent outbreaks of BSE.

Global temperatures reach their highest levels in 130 years.

The Ocean Dumping Ban legislates international dumping of wastes in the ocean.

EPA studies report that indoor air can be 100 times as polluted as outdoor air. Radon is found to be widespread in U.S. homes.

Beaches on the east coast of the United States are closed because of contamination by medical waste washed onshore.

The United States experiences its worst drought in 50 years.

Plastic ring six-pack holders are required to be made degradable.

U.S. legislation: Passage of the Plastic Pollution Research and Control Act bans ocean dumping of plastic materials.

1989 The United Kingdom bans the use of cattle brains, spinal cords, tonsils, thymuses, spleens, and intestines in foods intended for human consumption as a means of preventing further outbreaks of Creutzfeldt-Jakob disease (CJD), the human version of mad cow disease, in humans.

Fire aboard the *Kharg 5* releases 75,000 tons of oil into the sea surrounding the Canary Islands.

The Montreal Protocol treaty is updated and amended.

The New York Department of Environmental Conservation reports that 25 percent of the lakes and ponds in the Adirondacks are too acidic to support fish.

The *Exxon Valdez* runs aground on Prince William Sound, Alaska, spilling 11 million gallons of oil into one of the world's most fragile ecosystems.

1990 Ocean Robbins, age 16, and **Ryan Eliason**, 18, are the cofounders of YES!, or Youth for Environmental Sanity. The goal of YES! is to educate, inspire, and empower young people to take positive action for healthy people and a healthy planet. Robbins served as director for five years and is now the organization's president. As of 2000, the program has reached 600,000 students in 1,200 schools in 43 states through full school assemblies.

UN report forecasts a world temperature increase of 2°F within 35 years as a result of greenhouse gas emissions.

U.S. legislation: Passage of the Clean Air Act amendments including requirements to control the emission of sulfur dioxide and nitrogen oxides.

1991 The Gulf War concludes with hundreds of oil wells in Kuwait being set afire by Iraqi troops, resulting in extensive air and water pollution problems.

The United States accepts an agreement on Antarctica which prohibits activities relating to mining, protects native species of flora and fauna, and limits tourism and marine pollution.

Eight scientists begin a two-year stay in Biosphere 2 in Arizona, a test center designed to provide a self-sustaining habitat modeling Earth's natural environments. The experiment, which is repeated in 1993, meets with much criticism and is deemed largely unsuccessful.

1992 UN Earth Summit is held in Rio de Janeiro, Brazil. Major resolutions resulting from the summit include the Rio Declaration on Environment and Development, Agenda 21, Biodiversity Convention, Statement of Forest Principles, and the Global Warming Convention, which is signed by more than 160 nations.

Severn Cullis-Suzuki speaks for six minutes to the delegates urging them to work hard on resolving global environmental issues. She received a standing ovation.

The Montreal Protocol is again amended with signatories agreeing to phase out CFC use by the year 2000.

1993 Sugar producers and U.S. government agree on a restoration plan for the Florida Everglades.

1994 *Dumping in Dixie: Class and Environmental Quality* is published by **Robert Bullard**. The book reports on five environmental justice campaigns in states ranging from Texas to West Virginia. Bullard emphasizes that African Americans are concerned about and do participate in environmental issues.

The California Desert Protection Act is passed.

Failure of a dike results in the release of 102,000 tons of oil into the Siberian tundra near Usink in northern Russia.

The Russian government calls for preventive measures to control the destruction of Lake Baikail.

The bald eagle is reclassified from an endangered species to a threatened species on the U.S. Endangered Species List.

An 8.5-million-gallon spill is discovered in Unocal's Guadalupe oil field in California.

1995 The U.S. Government reintroduces endangered wolves to Yellowstone Park.

1999 Scientists report that the human population of Earth now exceeds 6 billion people.

The peregrine falcon is removed from the U.S. Endangered Species List.

The *New Carissa* runs aground off the coast of Oregon, leaking some oil into Coos Bay. The tanker is later towed into the open ocean and sunk.

Beyond Globalization: Shaping a Sustainable Global Economy is published by Hazel Henderson.

Paul Hawken coauthors *Natural Capitalism, Creating the Next Industrial Revolution.*

Off the Map, an Expedition Deep into Imperialism, the Global Economy, and Other Earthly Whereabouts is published by **Chellis Glendinning**.

Twenty-three-year-old **Julia Butterfly Hill** comes down out of a 180-foot California redwood tree after living there for two years to prevent the destruction of the forest. A deal is made with the logging company to spare the tree as well as a three-acre buffer zone.

2000 Denis Hayes is the coordinator and **Mark Dubois** is the international coordinator of Earthday 2000.

Ralph Nader and **Winona LaDuke** run for U.S. president and vice president on the Green Party ticket.

In January 2000, Hazel Wolf passes away at the age of 101.

The Chernobyl nuclear power plant is scheduled to close in December.

Anthropologists for the Wildlife Conservation Society in New York announce that a type of large West African monkey is extinct, making it the first primate to vanish in the twenty-first century.

A study by National Park Trust, a privately funded land conservancy, finds that more than 90,000 acres within state parks in 32 states are threatened by commercial and residential development and increased traffic, among other things.

A bone-dry summer in north-central Texas breaks the Depression-era drought record when the Dallas area marks 59 days without rain. The arid streak with 100-degree daily highs breaks a record of 58 days set in the midst of the Dust Bowl in 1934 and tied in 1950. The Texas drought exceeded 1 billion dollars in agricultural losses.

Massachusetts announces that the state will spend $600,000 to determine whether petroleum pollution in largely African American city neighborhoods contributes to lupus, a potentially deadly immune disease. The research, to be conducted over three years, will target three areas of the city with unusually high levels of petroleum contamination.

Hybrid vehicle Toyota Prius is offered for sale in the United States.

The hole in the ozone layer over Antarctica has stretched over a populated city for the first time, after ballooning to a new record size. Previously, the hole had opened only over Antarctica and the surrounding ocean.

2001 An environmental group that successfully campaigned for the return of wolves to Yellowstone National Park wants the federal government to do the same in western Colorado and parts of Utah, southern Wyoming, northern New Mexico, and Arizona.

The UN Environment Program launches a campaign to save the world's great apes from extinction, asking for at least $1 million to get started.

The captain and crew of a tanker that spilled at least 185,000 gallons of diesel into the fragile marine environment of the Galapagos Islands have been arrested.

One hundred sixty-five countries approve the Kyoto rules aimed at halting global warming. The Kyoto Protocol requires industrial countries to scale back emissions of carbon dioxide and other greenhouse gases by an average of 5 percent from their 1990 levels by 2012. The United States, the world's biggest polluter rejects the pact.

The EPA reaches an agreement for the phaseout of a widely used pesticide, diazinon, because of potential health risks to children.

For the second time in three years, the average fuel economy of new passenger cars and light trucks sold in the United States dropped to its lowest level since 1980.

More and more Americans are breathing dirtier air, and larger U.S. cities such as Los Angeles and Atlanta remain among the worst for pollution.

In rural stretches of Alaska, global warming has thinned the Arctic pack ice, making travel dangerous for native hunters. Traces of industrial pollution from distant continents is showing up in the fat of Alaska's marine wildlife and in the breast milk of native mothers who eat a traditional diet including seal and walrus meat.

2002 A Congo volcano devastates a Congolese town burning everything in its path, creating a five-foot-high wall of cooling stone, and leaving a half million people homeless.

New research is conducted in the practice of killing sharks solely for their fins.

A report by the USGS shows the nation's waterways are awash in traces of chemicals used in beauty aids, medications, cleaners, and foods. Among the substances are caffeine, painkillers, insect repellent, perfumes, and nicotine. These substances largely escape regulation and defy municipal wastewater treatment.

A microbe is discovered to be a major cause of the destruction of beech trees in the northeastern United States.

A study discovers that, if fallen leaves are left in stagnant water, they can release toxic mercury, which eventually can accumulate in fish that live far downstream.

Scientists are experimenting with various sprays containing clay particles to kill toxic algae in seawater.

Meteorologists discover that the Mediterranean Sea receives air current pollutants from Europe, Asia, and North America.

Researchers report possible ways of blocking the deadly effects of anthrax.

2003 A new international treaty—The Protocol on Persistent Organic Pollutants (POPS) was ratified by 17 nations although the United States has not signed on. The treaty drafted by United Nations reduces and eliminates 16 toxic chemicals that are long-lived in the environment and travel globally. The new treaty, an extension of an earlier one signed in 2000, added four more organic persistent pollutants to the list.

Many global scientific studies reveal that excessive ultraviolet (UV) sunlight and pollution are linked to a decline in amphibian populations. Now Canadian biologists find that too much exposure of excessive UV radiation to tadpole populations reduces their chances of becoming frogs.

2003 marked the 50th anniversary of the research and publication of a different structure of the DNA model proposed by James D. Watson and

Francis H.C. Crick. In 1953 the scientists reported that the DNA molecule resembled a spiral staircase.

A new excavation in South Africa discovered the oldest fossils in the human family. The bones of a skull and a partial arm found in two caves date back to 4 million years ago according to scientists in Johannesburg

Scientists in New Jersey discovered that some outdoor antimosquito coils used to keep insects away can also cause respiratory health problems. The spiral-shaped container releases pollutants in the fumes expelled from coil. The researchers suggest that consumers should check these products carefully.

Researchers in Australia reported that pieces of plastic litter found in oceans continue to have an effect on marine wildlife. Small plastic chips are a hazard for seabirds who mistake the litter for food or fish eggs. The litter also moves up the food chain from fish that have ingested the plastic chips and in turn seals eat them.

2004 A scientific study reported that consumers should limit their consumption of farm-raised Atlantic salmon because of high concentrations of chlorinated organic contaminants in the fish. Their study revealed that the farm-raised salmon were contaminated with polychlorinated biphenyls (PCBs) and other organic chemicals. Except for the PCBs, the researchers agree that the farm-raised fish are healthy but consumption should be limited to no more than once a month in the diet. The researchers based their dietary report on the U.S. Environmental Protection Agency cancer risk assessments.

A group in Salisbury Plain, England is restoring Stonehenge to its natural setting. As a popular historic site to visitors, Stonehenge had become an area surrounded by roads and parking lots. The new restoration plan calls for building an underground tunnel for traffic and removing one of the roads. The present parking lots will become open grassy lawns.

Experts reported that two billion people lack reliable access to safe and nutritious food and 800 million, 40 percent of them children, are classified as chronically malnourished.

Public health officials in Uganda have reported progress in the country's fight against HIV, the AIDS virus. Since 1990's HIV cases in Uganda have dropped by more than 60 percent. Unfortunately, Uganda's neighboring countries are not doing well in their HIV prevention programs.

United Nations Secretary-General Kofi Annan stated, "by 2025, two-thirds of the world's population may be living in countries that face serious water shortages." The growing population is making surface water scarcer particularly in urban areas.

United States and Israel scientists have found a way to produce hydrogen from water. The hydrogen energy can be used in making fuel cells to power vehicles and homes. The research team uses solar radiation to heat sodium hydroxide in a solution of water. At high temperatures the water molecules (H_2O) break apart into oxygen and hydrogen. Using solar-power to produce hydrogen is better environmentally than hydrogen derived from fossil fuels.

Appendix B: Endangered Species by State

The list below, obtained from the U.S. Fish and Wildlife Service, is an abridged listing of a selected group of endangered species (E) for each state. For a full list of endangered and threatened species, and other information about endangered species and the Endangered Species Act, see the Endangered Species Program Website at http://endangered.fws.gov/

ALABAMA

(Alabama has 106 plant and animal species that are listed as endangered (E) or threatened (T). The following list is only a selection of those plants and animals that are endangered. Contact the U.S. Fish and Wildlife Service to see the entire list.)

Animals

E - Bat, gray
E - Bat, Indiana
E - Cavefish, Alabama
T - Chub, spotfin
E - Clubshell, black
E - Combshell, southern
E - Darter, boulder
E - Fanshell
E - Kidneyshell, triangular
E - Lampmussel, Alabama
E - Manatee, West Indian
E - Moccasinshell, Coosa
E - Mouse, Alabama beach
E - Mussel, ring pink
E - Pearlymussel, cracking
E - Pearlymussel, Cumberland monkeyface
E - Pigtoe, dark
E - Plover, piping
E - Shrimp, Alabama cave
E - Snail, tulotoma (Alabama live-bearing)
E - Stork, wood
E - Turtle, Alabama redbelly (red-bellied)
E - Turtle, leatherback sea
E - Woodpecker, red-cockaded

Plants

E - Grass, Tennessee yellow-eyed
E - Leather-flower, Alabama
E - Morefield's leather-flower
E - Pinkroot, gentian
E - Pitcher-plant, Alabama canebrake
E - Pitcher-plant, green
E - Pondberry
E - Prairie-clover, leafy

ALASKA

Animals

E - Curlew, Eskimo (*Numenius borealis*)
E - Falcon, American peregrine (*Falco peregrinus anatum*)

Plant

E - Aleutian shield-fern (Aleutian holly-fern) (*Polystichum aleuticum*)

ARIZONA

Animals

E - Ambersnail, Kanab
E - Bat, lesser (Sanborn's) long-nosed
E - Bobwhite, masked (quail)
E - Chub, bonytail
E - Chub, humpback
E - Chub, Virgin River
E - Chub, Yaqui
E - Flycatcher, Southwestern willow
E - Jaguarundi
E - Ocelot
E - Pronghorn, Sonoran
E - Pupfish, desert
E - Rail, Yuma clapper
E - Squawfish, Colorado
E - Squirrel, Mount Graham red
E - Sucker, razorback
E - Topminnow, Gila (incl. Yaqui)
E - Trout, Gila
E - Vole, Hualapai Mexican
E - Woundfin

Plants

E - Arizona agave
E - Arizona cliffrose
E - Arizona hedgehog cactus
E - Brady pincushion cactus
E - Kearney's blue-star
E - Nichol's Turk's head cactus
E - Peebles Navajo cactus
E - Pima pineapple cactus
E - Sentry milk-vetch

ARKANSAS

Animals

E - Bat, gray
E - Bat, Indiana
E - Bat, Ozark big-eared
E - Beetle, American burying (giant carrion)
E - Crayfish, cave
E - Pearlymussel, Curtis'
E - Pearlymussel, pink mucket
E - Pocketbook, fat
E - Pocketbook, speckled
E - Rock-pocketbook, Ouachita (Wheeler's pearly mussel)
E - Sturgeon, pallid
E - Tern, least
E - Woodpecker, red-cockaded

Plants

E - Harperella
E - Pondberry
E - Running buffalo clover

CALIFORNIA

(California has more than 160 plant and animal species that are listed as endangered or threatened. The following list is only a selection of those plants and animals that are endangered. Contact the U.S. Fish and Wildlife Service to see the entire list.)

Animals

E - Butterfly, El Segundo blue
E - Butterfly, Lange's metalmark
E - Chub, Mohave tui
E - Condor, California
E - Crayfish, Shasta (placid)
E - Fairy shrimp, Conservancy
E - Falcon, American peregrine
E - Fly, Delhi Sands flower-loving
E - Flycatcher, Southwestern willow
E - Fox, San Joaquin kit
E - Goby, tidewater
E - Kangaroo rat, Fresno
E - Lizard, blunt-nosed leopard
E - Mountain beaver, Point Arena
E - Mouse, Pacific pocket
E - Pelican, brown
E - Pupfish, Owens
E - Rail, California clapper
E - Salamander, Santa Cruz long-toed
E - Shrike, San Clemente loggerhead
E - Shrimp, California freshwater
E - Snail, Morro shoulderband (banded dune)
E - Snake, San Francisco garter
E - Stickleback, unarmored threespine
E - Sucker, Lost River
E - Tadpole shrimp, vernal pool
E - Tern, California least
E - Toad, arroyo southwestern
E - Turtle, leatherback sea
E - Vireo, least Bell's
E - Vole, Amargosa

Plants

E - Antioch Dunes evening-primrose
E - Bakersfield cactus
E - Ben Lomond wallflower
E - Burke's goldfields
E - California jewelflower
E - California Orcutt grass
E - Clover lupine
E - Cushenbury buckwheat
E - Fountain thistle
E - Gambel's watercress
E - Kern mallow
E - Loch Lomond coyote-thistle
E - Robust spineflower (includes Scotts Valley spineflower)
E - San Clemente Island larkspur
E - San Diego button-celery
E - San Mateo thornmint
E - Santa Ana River woolly-star
E - Santa Barbara Island liveforever
E - Santa Cruz cypress
E - Solano grass
E - Sonoma sunshine (Baker's stickyseed)

E - Stebbins' morning-glory
E - Truckee barberry
E - Western lily

COLORADO

Animals

E - Butterfly, Uncompahgre fritillary
E - Chub, bonytail
E - Chub, humpback
E - Crane, whooping
E - Ferret, black-footed
E - Flycatcher, Southwestern willow
E - Plover, piping
E - Squawfish, Colorado
E - Sucker, razorback
E - Tern, least
E - Wolf, gray

Plants

E - Clay-loving wild-buckwheat
E - Knowlton cactus
E - Mancos milk-vetch
E - North Park phacelia
E - Osterhout milk-vetch
E - Penland beardtongue

CONNECTICUT

Animals

E - Mussel, dwarf wedge
E - Plover, piping
E - Tern, roseate
E - Turtle, hawksbill sea
E - Turtle, Kemp's (Atlantic) ridley sea
E - Turtle, leatherback sea

Plant

E - Sandplain gerardia

DELAWARE

Animals

E - Plover, piping
E - Squirrel, Delmarva Peninsula fox
E - Turtle, hawksbill sea
E - Turtle, Kemp's (Atlantic) ridley sea—Turtle, green sea
E - Turtle, leatherback sea

Plant

E - Canby's dropwort

FLORIDA

(Florida has more than 90 plant and animal species that are listed as endangered or threatened. The following list is only a selection of those plants and animals that are endangered. Contact the U.S. Fish and Wildlife Service to see the entire list.)

Animals

E - Bat, gray
E - Butterfly, Schaus swallowtail
E - Crocodile, American
E - Darter, Okaloosa
E - Deer, key
E - Kite, Everglade snail
E - Manatee, West Indian (Florida)
E - Mouse, Anastasia Island beach
E - Mouse, Choctawahatchee beach
E - Panther, Florida
E - Plover, piping
E - Rabbit, Lower Keys
E - Rice rat (silver rice rat)
E - Sparrow, Cape Sable seaside
E - Stork, wood
E - Tern, roseate
E - Turtle, hawksbill sea
E - Turtle, Kemp's (Atlantic) ridley sea
E - Turtle, leatherback sea
E - Vole, Florida salt marsh
E - Woodpecker, red-cockaded
E - Woodrat, Key Largo

Plants

E - Apalachicola rosemary
E - Beautiful pawpaw
E - Brooksville (Robins') bellflower
E - Carter's mustard
E - Chapman rhododendron
E - Cooley's water-willow
E - Crenulate lead-plant
E - Etonia rosemary
E - Florida golden aster
E - Fragrant prickly-apple
E - Garrett's mint
E - Key tree-cactus
E - Lakela's mint
E - Okeechobee gourd

E - Scrub blazingstar
E - Small's milkpea
E - Snakeroot
E - Wireweed

GEORGIA

Animals

E - Acornshell, southern
E - Bat, gray
E - Bat, Indiana
E - Clubshell, ovate
E - Clubshell, southern
E - Combshell, upland
E - Darter, amber
E - Darter, Etowah
E - Kidneyshell, triangular
E - Logperch, Conasauga
E - Manatee, West Indian (Florida)
E - Moccasinshell, Coosa
E - Pigtoe, southern
E - Plover, piping
E - Stork, wood
E - Turtle, hawksbill sea
E - Turtle, Kemp's (Atlantic) ridley sea
E - Turtle, leatherback sea
E - Woodpecker, red-cockaded

Plants

E - American chaffseed
E - Black-spored quillwort
E - Canby's dropwort
E - Florida torreya
E - Fringed campion
E - Green pitcher-plant
E - Hairy rattleweed
E - Harperella
E - Large-flowered skullcap
E - Mat-forming quillwort
E - Michaux's sumac
E - Persistent trillium
E - Pondberry
E - Relict trillium
E - Smooth coneflower
E - Tennessee yellow-eyed grass

HAWAII

(Hawaii has 300 plant and animal species listed as endangered or threatened. The following list is only a selection of those plants and animals that are endangered. Contact the U.S. Fish and Wildlife Service to see the entire list.)

Animals

E - 'Akepa, Hawaii (honeycreeper)
E - Bat, Hawaiian hoary
E - Coot, Hawaiian
E - Creeper, Hawaiian
E - Crow, Hawaiian
E - Duck, Hawaiian
E - Duck, Laysan
E - Finch, Laysan (honeycreeper)
E - Finch, Nihoa (honeycreeper)
E - Goose, Hawaiian (nene)
E - Hawk, Hawaiian
E - Millerbird, Nihoa (old world warbler)
E - Nukupu'u (honeycreeper)
E - Palila (honeycreeper)
E - Parrotbill, Maui (honeycreeper)
E - Petrel, Hawaiian dark-rumped
E - Snails, Oahu tree
E - Stilt, Hawaiian
E - Turtle, hawksbill sea
E - Turtle, leatherback sea

Plants

E - Abutilon eremitopetalum
E - Bonamia menziesii
E - Carter's panicgrass
E - Diamond Head schiedea
E - Dwarf iliau
E - Fosberg's love grass
E - Hawaiian bluegrass
E - Hawaiian red-flowered geranium
E - Kaulu
E - Kiponapona
E - Mahoe
E - Mapele
E - Nanu
E - Nehe
E - Opuhe
E - Pamakani
E - Round-leaved chaff-flower
E - Viola helenae

IDAHO

Animals

E - Caribou, woodland
E - Crane, whooping

E - Limpet, Banbury Springs
E - Snail, Snake River physa
E - Snail, Utah valvata
E - Springsnail, Bruneau Hot
E - Springsnail, Idaho
E - Sturgeon, white
E - Wolf, gray

Plants

(No plants on the endangered list)

ILLINOIS

Animals

E - Bat, gray
E - Bat, Indiana
E - Butterfly, Karner blue
E - Dragonfly, Hine's emerald
E - Falcon, American peregrine
E - Fanshell
E - Pearlymussel, Higgins' eye
E - Pearlymussel, orange-foot pimple back
E - Pearlymussel, pink mucket
E,T - Plover, piping
E - Pocketbook, fat
E - Snail, Iowa Pleistocene
E - Sturgeon, pallid
E - Tern, least

Plant

E - Leafy prairie-clover

INDIANA

Animals

E - Bat, gray
E - Bat, Indiana
E - Butterfly, Karner blue
E - Butterfly, Mitchell's satyr
E - Clubshell
E - Fanshell
E - Mussel, ring pink (golf stick pearly)
E - Pearlymussel, cracking
E - Pearlymussel, orange-foot pimple back
E - Pearlymussel, pink mucket
E - Pearlymussel, tubercled-blossom
E - Pearlymussel, white cat's paw
E - Pearlymussel, white wartyback
E - Pigtoe, rough
E,T - Plover, piping
E - Pocketbook, fat
E - Riffleshell, northern
E - Tern, least

Plant

E - Running buffalo clover

IOWA

Animals

E - Bat, Indiana
E - Pearlymussel, Higgins' eye
E - Plover, piping
E - Snail, Iowa Pleistocene
E - Sturgeon, pallid
E - Tern, least

Plants

(No plants on the endangered list)

KANSAS

Animals

E - Bat, gray
E - Bat, Indiana
E - Crane, whooping
E - Curlew, Eskimo
E - Ferret, black-footed
E - Plover, piping
E - Sturgeon, pallid
E - Tern, least
E - Vireo, black-capped

Plants

(No plants on endangered list)

KENTUCKY

Animals

E - Bat, gray
E - Bat, Indiana
E - Bat, Virginia big-eared
E - Clubshell
E - Darter, relict
E - Falcon, American peregrine
E - Fanshell
E - Mussel, ring pink (golf stick pearly)
E - Mussel, winged mapleleaf
E - Pearlymussel, cracking
E - Pearlymussel, Cumberland bean

E - Pearlymussel, dromedary
E - Pearlymussel, little-wing
E - Pearlymussel, orange-foot pimple back
E - Pearlymussel, pink mucket
E - Pearlymussel, purple cat's paw
E - Pearlymussel, tubercled-blossom
E - Pearlymussel, white wartyback
E - Pigtoe, rough
E - Plover, piping
E - Pocketbook, fat
E - Riffleshell, northern
E - Riffleshell, tan
E - Shiner, Palezone
E - Shrimp, Kentucky cave
E - Sturgeon, pallid
E - Tern, least
E - Woodpecker, red-cockaded

Plants

E - Cumberland sandwort
E - Rock cress
E - Running buffalo clover
E - Short's goldenrod

LOUISIANA

Animals

E - Manatee, West Indian (Florida)
E - Pearlymussel, pink mucket
E - Pelican, brown
E - Plover, piping
E - Sturgeon, pallid
E - Tern, least
T - Turtle, green sea
E - Turtle, hawksbill sea
E - Turtle, Kemp's (Atlantic) ridley sea
E - Turtle, leatherback sea
E - Vireo, black-capped
E - Woodpecker, red-cockaded

Plants

E - American chaffseed
E - Louisiana quillwort
E - Pondberry

MAINE

Animals

E - Plover, piping
E - Tern, roseate
E - Turtle, leatherback sea

Plant

E - Furbish lousewort

MARYLAND

Animals

E - Bat, Indiana
E - Darter, Maryland
E - Mussel, dwarf wedge
E - Plover, piping
E - Squirrel, Delmarva Peninsula fox
E - Turtle, hawksbill sea
E - Turtle, Kemp's (Atlantic) ridley sea
E - Turtle, leatherback sea

Plants

E - Canby's dropwort
E - Harperella
E - Northeastern (Barbed bristle) bulrush
E - Sandplain gerardia

MASSACHUSETTS

Animals

E - Beetle, American burying (giant carrion)
E - Falcon, American peregrine
E - Mussel, dwarf wedge
E - Plover, piping
E - Tern, roseate
E - Turtle, hawksbill sea
E - Turtle, Kemp's (Atlantic) ridley sea
E - Turtle, leatherback sea
E - Turtle, Plymouth redbelly (red-bellied)

Plants

E - Northeastern (Barbed bristle)
E - Sandplain gerardia

MICHIGAN

Animals

E - Bat, Indiana
E - Beetle, American burying (giant carrion)
E - Beetle, Hungerford's crawling water
E - Butterfly, Karner blue
E - Butterfly, Mitchell's satyr
E - Clubshell
E - Plover, piping
E - Riffleshell, northern
E - Warbler, Kirtland's
E - Wolf, gray

Plant

E - Michigan monkey-flower

MINNESOTA

Animals

E - Butterfly, Karner blue
E - Mussel, winged mapleleaf
E - Pearlymussel, Higgins' eye
E - Plover, piping
E - Wolf, gray

Plant

E - Minnesota trout lily

MISSISSIPPI

Animals

E - Bat, Indiana
E - Clubshell, black (Curtus' mussel)
E - Clubshell, ovate
E - Clubshell, southern
E - Combshell, southern (penitent mussel)
E - Crane, Mississippi sandhill
E - Falcon, American peregrine
E - Manatee, West Indian (Florida)
E - Pelican, brown
E - Pigtoe, flat (Marshall's mussel)
E - Pigtoe, heavy (Judge Tait's mussel)
E - Plover, piping
E - Pocketbook, fat
E - Stirrupshell
E - Sturgeon, pallid
E - Tern, least
E - Turtle, hawksbill sea
E - Turtle, Kemp's (Atlantic) ridley sea
E - Turtle, leatherback sea
E - Woodpecker, red-cockaded

Plants

E - American chaffseed
E - Pondberry

MISSOURI

Animals

E - Bat, gray
E - Bat, Indiana
E - Bat, Ozark big-eared
E - Pearlymussel, Curtis'
E - Pearlymussel, Higgins' eye
E - Pearlymussel, pink mucket
E - Plover, piping
E - Pocketbook, fat
E - Sturgeon, pallid
E - Tern, least

Plants

E - Missouri bladderpod
E - Pondberry
E - Running buffalo clover

MONTANA

Animals

E - Crane, whooping
E - Curlew, Eskimo
E - Ferret, black-footed
E - Plover, piping
E - Sturgeon, pallid
E - Sturgeon, white
E - Tern, least
E - Wolf, gray

Plants

(No plants on endangered list)

NEBRASKA

Animals

E - Beetle, American burying (giant carrion)
E - Crane, whooping
E - Curlew, Eskimo
E - Ferret, black-footed
E - Plover, piping
E - Sturgeon, pallid
E - Tern, least

Plant

E - Blowout penstemon

NEVADA

Animals

E - Chub, bonytail
E - Chub, Pahranagat roundtail (bonytail)
E - Chub, Virgin River
E - Cui-ui
E - Dace, Ash Meadows speckled

E - Dace, Clover Valley speckled
E - Dace, Independence Valley speckled
E - Dace, Moapa
E - Poolfish (killifish), Pahrump
E - Pupfish, Ash Meadows Amargosa
E - Pupfish, Devils Hole
E - Pupfish, Warm Springs
E - Spinedace, White River
E - Springfish, Hiko White River
E - Springfish, White River
E - Sucker, razorback
E - Woundfin

Plants

E - Amargosa niterwort
E - Steamboat buckwheat

NEW HAMPSHIRE

Animals

E - Butterfly, Karner blue
E - Mussel, dwarf wedge
E - Turtle, leatherback sea

Plants

E - Jesup's milk-vetch
E - Northeastern (Barbed bristle) bulrush
E - Robbins' cinquefoil

NEW JERSEY

Animals

E - Bat, Indiana
E - Plover, piping
E - Tern, roseate
E - Turtle, hawksbill sea
E - Turtle, Kemp's (Atlantic) ridley sea
E - Turtle, leatherback sea

Plant

E - American chaffseed

NEW MEXICO

Animals

E - Bat, lesser (Sanborn's) long-nosed
E - Bat, Mexican long-nosed
E - Crane, whooping
E - Gambusia, Pecos
E - Isopod, Socorro
E - Minnow, Rio Grande silvery
E - Springsnail, Alamosa
E - Springsnail, Socorro
E - Sucker, razorback
E - Tern, least
E - Topminnow, Gila (incl. Yaqui)
E - Trout, Gila
E - Woundfin

Plants

E - Holy Ghost ipomopsis
E - Knowlton cactus
E - Kuenzler hedgehog cactus
E - Lloyd's hedgehog cactus
E - Mancos milk-vetch
E - Sacramento prickly-poppy
E - Sneed pincushion cactus
E - Todsen's pennyroyal

NEW YORK

Animals

E - Butterfly, Karner blue
E - Mussel, dwarf wedge
E, T - Plover, piping
E - Tern, roseate
E - Turtle, hawksbill sea
E - Turtle, Kemp's (Atlantic) ridley sea
E - Turtle, leatherback sea

Plants

E - Northeastern (Barbed bristle) bulrush
E - Sandplain gerardia

NORTH CAROLINA

Animals

E - Bat, Indiana
E - Bat, Virginia big-eared
E - Butterfly, Saint Francis' satyr
E - Elktoe, Appalachian
E - Falcon, American peregrine
E - Heelsplitter, Carolina
E - Manatee, West Indian (Florida)
E - Mussel, dwarf wedge
E - Pearlymussel, little-wing
E - Plover, piping
E - Shiner, Cape Fear
E - Spider, spruce-fir moss

E - Spinymussel, Tar River
E - Squirrel, Carolina northern flying
E - Tern, roseate
E - Turtle, hawksbill sea
E - Turtle, Kemp's (Atlantic) ridley sea
E - Turtle, leatherback sea
E - Wolf, red
E - Woodpecker, red-cockaded

Plants

E - American chaffseed
E - Bunched arrowhead
E - Canby's dropwort
E - Cooley's meadowrue
E - Green pitcher-plant
E - Harperella
E - Michaux's sumac
E - Mountain sweet pitcher-plant
E - Pondberry
E - Roan Mountain bluet
E - Rock gnome lichen
E - Rough-leaved loosestrife
E - Schweinitz's sunflower
E - Small-anthered bittercress
E - Smooth coneflower
E - Spreading avens
E - White irisette

NORTH DAKOTA

Animals

E - Crane, whooping
E - Curlew, Eskimo
E - Falcon, American peregrine
E - Ferret, black-footed
E - Plover, piping
E - Sturgeon, pallid
E - Tern, least
E - Wolf, gray

Plants

(No plants on endangered list)

OHIO

Animals

E - Bat, Indiana
E - Beetle, American burying (giant carrion)
E - Butterfly, Karner blue
E - Butterfly, Mitchell's satyr
E - Clubshell
E - Dragonfly, Hine's emerald
E - Fanshell
E - Madtom, Scioto
E - Pearlymussel, pink mucket
E - Pearlymussel, purple cat's paw
E - Pearlymussel, white cat's paw
E, T - Plover, piping
E - Riffleshell, northern

Plant

E - Running buffalo clover

OKLAHOMA

Animals

E - Bat, gray
E - Bat, Indiana
E - Bat, Ozark big-eared
E - Beetle, American burying (giant carrion)
E - Crane, whooping
E - Curlew, Eskimo
E - Plover, piping
E - Rock-pocketbook, Ouachita
E - Tern, least
E - Vireo, black-capped
E - Woodpecker, red-cockaded

Plants

(No plants on the endangered list)

OREGON

Animals

E - Chub, Borax Lake
E - Chub, Oregon
E - Deer, Columbian white-tailed
E - Pelican, brown
E - Sucker, Lost River
E - Sucker, shortnose
E - Turtle, leatherback sea

Plants

E - Applegate's milk-vetch
E - Bradshaw's desert-parsley
E - Malheur wire-lettuce
E - Marsh sandwort
E - Western lily

PENNSYLVANIA

Animals

E - Bat, Indiana
E - Clubshell
E - Mussel, dwarf wedge
E - Mussel, ring pink (golf stick pearly)
E - Pearlymussel, cracking
E - Pearlymussel, orange-foot pimple back
E - Pearlymussel, pink mucket
E - Pigtoe, rough
E,T - Plover, piping
E - Riffleshell, northern

Plant

E - Northeastern (Barbed bristle) bulrush

RHODE ISLAND

Animals

E - Beetle, American burying
E - Falcon, American peregrine
E - Plover, piping
E - Tern, roseate
E - Turtle, hawksbill sea
E - Turtle, Kemp's
E - Turtle, leatherback sea

Plant

E - Sandplain gerardia

SOUTH CAROLINA

Animals

E - Bat, Indiana
E - Heelsplitter, Carolina
E - Manatee, West Indian (Florida)
E - Plover, piping
E - Stork, wood
E - Tern, roseate
E - Turtle, hawksbill sea
E - Turtle, Kemp's (Atlantic) ridley sea
E - Turtle, leatherback sea
E - Woodpecker, red-cockaded

Plants

E - American chaffseed
E - Black-spored quillwort
E - Bunched arrowhead
E - Canby's dropwort
T - Dwarf-flowered heartleaf
E - Harperella
E - Michaux's sumac
E - Mountain sweet pitcher-plant
E - Persistent trillium
E - Pondberry
E - Relict trillium
E - Rough-leaved loosestrife
E - Schweinitz's sunflower
E - Smooth coneflower

SOUTH DAKOTA

Animals

E - Beetle, American burying (giant carrion)
E - Crane, whooping
E - Curlew, Eskimo
E - Ferret, black-footed
E - Plover, piping
E - Sturgeon, pallid
E - Tern, least
E - Wolf, gray

Plants

(No plants on the endangered list)

TENNESSEE

(Tennessee has 81 plant and animal species that are listed as endangered or threatened. The following list is only a selection of those plants and animals that are endangered. Contact the U.S. Fish and Wildlife Service to see the entire list.)

Animals

E - Bat, gray
E - Bat, Indiana
E - Combshell, upland
E - Crayfish, Nashville
E - Darter, amber
E - Fanshell
E - Lampmussel, Alabama
E - Madtom, Smoky
E - Marstonia (snail), (royalobese)
E - Moccasinshell, Coosa
E - Mussel, ring pink (golf stick pearly)
E - Pearlymussel, Appalachian monkeyface

E - Pearlymussel, Cumberland bean
E - Riversnail, Anthony's
E - Spider, spruce-fir moss
E - Squirrel, Carolina northern flying
E - Sturgeon, pallid
E - Tern, least
E - Wolf, red
E - Woodpecker, red-cockaded

Plants

E - Cumberland sandwort
E - Green pitcher-plant
E - Large-flowered skullcap
E - Leafy prairie-clover (Dalea)
E - Roan Mountain bluet
E - Rock cress
E - Rock gnome lichen
E - Ruth's golden aster
E - Spring Creek bladderpod
E - Tennessee purple coneflower
E - Tennessee yellow-eyed grass

TEXAS

(Texas has 70 plant and animal species that are listed as endangered or threatened. The following list is only a selection of those plants and animals that are endangered. Contact the U.S. Fish and Wildlife Service to see the entire list.)

Animals

E - Bat, Mexican long-nosed
E - Beetle, Coffin Cave mold
E - Crane, whooping
E - Curlew, Eskimo
E - Darter, fountain
E - Falcon, northern aplomado
E - Jaguarundi
E - Manatee, West Indian (Florida)
E - Minnow, Rio Grande silvery
E - Ocelot
E - Pelican, brown
E - Plover, piping
E - Prairie-chicken, Attwater's greater
E - Pupfish, Comanche Springs
E - Salamander, Texas blind
E - Spider, Tooth Cave
E - Tern, least
E - Toad, Houston
E - Turtle, hawksbill sea
E - Turtle, Kemp's (Atlantic) ridley sea
E - Vireo, black-capped
E - Warbler, golden-cheeked
E - Woodpecker, red-cockaded

Plants

E - Ashy dogweed
E - Black lace cactus
T - Hinckley's oak
E - Large-fruited sand-verbena
E - Little Aguja pondweed
E - Lloyd's hedgehog cactus
E - Nellie cory cactus
E - Sneed pincushion cactus
E - South Texas ambrosia
E - Star cactus
E - Terlingua Creek cats-eye
E - Texas poppy-mallow
E - Texas snowbells
E - Texas wild-rice
E - Tobusch fishhook cactus
E - Walker's manioc

UTAH

Animals

E - Ambersnail, Kanab
E - Chub, bonytail
E - Chub, humpback
E - Chub, Virgin River
E - Crane, whooping
E - Ferret, black-footed
E - Flycatcher, Southwestern willow
E - Snail, Utah valvata
E - Squawfish, Colorado
E - Sucker, June
E - Sucker, razorback
E - Woundfin

Plants

E - Autumn buttercup
E - Barneby reed-mustard
E - Barneby ridge-cress (peppercress)
E - Clay phacelia
E - Dwarf bear-poppy
E - Kodachrome bladderpod
E - San Rafael cactus
E - Shrubby reed-mustard (toad-flax cress)
E - Wright fishhook cactus

VERMONT

Animals

E - Bat, Indiana
E - Mussel, dwarf wedge

Plants

E - Jesup's milk-vetch
E - Northeastern (Barbed bristle) bulrush

VIRGINIA

Animals

E - Bat, gray
E - Bat, Indiana
E - Bat, Virginia big-eared
E - Darter, duskytail
E - Falcon, American peregrine
E - Fanshell
E - Isopod, Lee County cave
E - Logperch, Roanoke
E - Mussel, dwarf wedge
E - Pearlymussel, Appalachian monkeyface
E - Pearlymussel, birdwing
E - Pearlymussel, cracking
E - Pearlymussel, Cumberland monkeyface
E - Pearlymussel, dromedary
E - Pearlymussel, green-blossom
E - Pearlymussel, little-wing
E - Pearlymussel, pink mucket
E - Pigtoe, fine-rayed
E - Pigtoe, rough
E - Pigtoe, shiny
E - Plover, piping
E - Riffleshell, tan
E - Salamander, Shenandoah
E - Snail, Virginia fringed mountain
E - Spinymussel, James River (Virginia)
E - Squirrel, Delmarva Peninsula fox
E - Squirrel, Virginia northern flying
E - Turtle, hawksbill sea
E - Turtle, Kemp's (Atlantic) ridley sea
E - Turtle, leatherback sea
E - Woodpecker, red-cockaded

Plants

E - Northeastern (Barbed bristle) bulrush
E - Peter's Mountain mallow
E - Shale barren rock-cress
E - Smooth coneflower

WASHINGTON

Animals

E - Caribou, woodland
E - Deer, Columbian white-tailed
E - Pelican, brown
E - Turtle, leatherback sea
E - Wolf, gray

Plants

E - Bradshaw's desert-parsley (lomatium)
E - Marsh sandwort

WEST VIRGINIA

Animals

E - Bat, Indiana
E - Bat, Virginia big-eared
E - Clubshell
E - Falcon, American peregrine
E - Fanshell
E - Mussel, ring pink
E - Pearlymussel, pink mucket
E - Pearlymussel, tubercled-blossom
E - Riffleshell, northern
E - Spinymussel, James River
E - Squirrel, Virginia northern flying

Plants

E - Harperella
E - Northeastern (Barbed bristle) bulrush
E - Running buffalo clover
E - Shale barren rock-cress

WISCONSIN

Animals

E - Butterfly, Karner blue
E - Dragonfly, Hine's emerald
E - Mussel, winged mapleleaf
E - Pearlymussel, Higgins' eye
E, T - Plover, piping
E - Warbler, Kirtland's
E - Wolf, gray

Plants

(No plants on endangered list)

WYOMING

Animals

E - Crane, whooping
E - Dace, Kendall Warm Springs
E - Ferret, black-footed
E - Squawfish, Colorado
E - Sucker, razorback
E - Toad, Wyoming
E - Wolf, gray

Plants

(No plants on endangered list)

Appendix C: Websites by Classification

Please note that the authors have made a consistent effort to include up-to-date Websites. However, over time, some Websites may move or no longer be posted.

ACID MINE DRAINAGE

National Reclamation Center, West Virginia University, Evansdale office, http//www.nrcce.wvu.edu.

ACID RAIN

http://www.epa.gov/docs/acidrain/andhome/html.

The EPA has a hotline to request educational materials or respond to questions regarding acid rain: (202) 343–9620. http://www.econet.apc.org/acid rain.

Environmental Protection Agency, http://www.epa.gov/docs/acidrain/effects/enveffct.html.

National Reclamation Center's West Virginia University, Evansdale office: http://www.nrcce.wvu.edu/

USGS Water Science/Acid Rain, http://wwwga.usgs.gov/edu/acidrain.html.

AGENCY FOR TOXIC SUBSTANCES AND DISEASES

Registry Division of Toxicology
1600 Clifton Road NE Mailstop E-29
Atlanta, GA 30333
Website: http://www.atsdr1.atsdr.cdc.gov:8080/atsdrhome.html.

Agency for Toxic Substances and Disease Registry, http://www.atsdr.cdc.gov/cxcx3.html.

Information on biosphere reserves and UNESCO's Man and the Biosphere Programme, UNESCO: http://www.unesco. org

Man and the Biosphere Program: http://www. mabnet.org

AGRICULTURE

United States Department of Agriculture, http://www.usda.gov.

ALTERNATIVE FUELS

Department of Energy, http://www.doe.gov.

Department of Energy Alternative Fuels Data Center, http://www.afdc.nrel.gov; http://www.afdc.doe.gov/; or http://www.fleets.doe.gov.

AMPHIBIANS

http://www.frogweb.gov/

ANTARCTICA

Antarctica Treaty, http://www.sedac.ciesin.org/pidb/register/reg-024.rrr.html.

Greenpeace International Antarctic Homepage, http://www.greenpeace.org/~comms/98/antarctic.

International Centre for Antarctic Information and Research Homepage (includes text of Antarctic Treaty), http://www.icair.iac.org.nz.

Virtual Antarctica, http://www.exploratorium.edu

ARCTIC

Arctic Circle (University of Connecticut), http://arcticcircle.uconn.edu/arcticcircle.

Arctic Council Home Page, http://www.nrc.ca/arctic/index.html.

Arctic Monitoring and Assessment Programme (Norway), http://www.gsf.de/ UNEP/amap1.html.

Arctic National Wildlife Refuge, http://energy.usgs.gov/factsheets/ANWR/ANWR.html.

Institute of Arctic and Alpine Research, http://instaar.colorado.edu.

Institute of the North (Alaska Pacific University),

Inuit Circumpolar Conference,
NOAA Fisheries, http://www.nmfs.gov/.

Nunavut,
Smithsonian Institution Arctic Studies Center, http://www.mnh.si.edu/arctic.

U.S. Fish and Wildlife Service
U.S. Department of the Interior

1849 C Street, NW,
Washington, D.C. 20240
Telephone: (202) 208-5634
Website: http://www.fws.gov.

World Conservation Monitoring Centre Arctic Programme, http://www.wcmc.org.uk/arctic.

AUTOMOBILE

Cars and Their Enviromental Impact, http://www.environment.volvocars.com/ch1-1.htm.

National Center for Vehicle Emissions Control and Safety (NCVECS), http://www.colostate.edu/Depts/NCVECS/ncvecs1.html.

U.S. Environmental Protection Agency Fact Sheet (EPA 400-F-92-004, August 1994), "Air Toxics from Motor Vehicles," http://www.epa.gov/oms/02-toxic.htm.

U.S. Enviromental Protection Agency, Office of Mobile Sources, http://www.epa.gov/oms.

BIOLOGICAL WEAPONS

Federation of American Scientist Biological Weapons Control, http://www.fas.org/bwc.

Chemical and Biological Defense Information Analysis Center, http://www.cbiac.apgea.army. mil

BIOMES

Committee for the National Institute for the Environment, http://www.cnie.org/nle/biodv-6.html.

BIOREMEDIATION

Consortium, http://www.rtdf.org/public/biorem.

BROWNFIELD

Projects, http://www.epa.gov/brownfields/.

CERES

Website: http://www.ceres.org or e-mail ceres@igc.apc.org.

Summaries of Major Environmental Laws, http://www.epa.gov/region5/defs/index.html.

CHEETAHS

Cheetah Conservation Fund

4649 Sunnyside Avenue N, Suite 325
Seattle, WA 98103
Website: http://www.cheetah.org.

World Wildlife Fund

1250 24th Street, NW,
Washington, D.C. 20037
Telephone: 1-800-225-5993
Website: http://www.worldwildlife.org/.

CHEMICAL WEAPONS

Chemical Stockpile Disposal Project (CSDP), http://www.pmcd.apgea.army.mil/graphical/CSDP/index.html.

Tooele Chemical Agent Disposal Site Facility, http://www.deq.state.ut.us/eqshw/cds/tocdfhp1.htm.

CLEAN WATER ACT

Sierra Club, "Happy 25th Birthday, Clean Water Act," http://sierraclub.org/wetlands/cwabday.html.

CLIMATE CHANGE AND GLOBAL WARMING

U.S. Geological Survey, Climate Change and History, http://geology.usgs.gov/index.shtml.

EPA Global Warming Site, http://www.epa.gov/globalwarming.

Greenpeace International, Climate, http://www.greenpeace.org/~climate.

United Nations Intergovernmental Panel on Climate Change, http://www.ipcc.ch.

COAL

Coal Age Magazine, http://coalage.com.

Department of Energy, Office of Fossil Energy, http:/www.doe.gov.

U.S. Geological Survey, National Coal Resources Data System, http:energy.er.usgs.gov/coalqual. htm.

COASTAL AND MARINE GEOLOGY

U.S. Geological Survey, http://marine.usgs.gov/.

COMPOSTING

EPA Office of Solid Waste and Emergency Response—Composting, http:www.epa.gov/epaoswer/non-hw/compost/index.htm

Cornell Composting, http://www.cfe.cornell.edu/compost/Composting_Homepage.html

CONSENT DECREES

EPA Office of Enforcement and Compliance Assurance, http://es.epa.gov/oeca/osre/decree.html.

CORAL REEFS

Coral Reef Alliance, http://www.coral.org.

Coral Reef Network Directory, Greenpeace
1436 U Street, NW
Washington, D.C. 20009
Website: http://www.greenpeace.org.

EARTHDAY 2000

Earth Day Network

91 Marion Street,
Seattle, WA 98104
Telephone: 1(206)-264-0114.
Website: http://www.earthday.net/; and worldwide@earthday.net.

EARTHWATCH

Earthwatch Institute International, http://www. earthwatch.org.

EL NIÑO

El Niño/La Niña theme page, contact NOAA Website: http://www.pmel.noaa.gov/toga-tao/el-nino/nino-home-low.html.

NOAA, La Niña homepage, www.elnino.noaa.gov/lanina.html.

National Center for Atmospheric Research, http://www.ncar.ucar.edu/.

National Hurricane Center/Tropical Prediction Center, http://www.nhc.noaa.gov/.

National Oceanographic and Atmospheric Administration, http://www.noaa.gov/.

Scripps Institute of Oceanography, http://sio.ucsd.edu/supp_groups/siocomm/elnino/elnino.html.

ELECTRIC VEHICLES

Electric Vehicle Association of the Americas 800-438-3228, http://www.evaa.org.

Electric Vehicle Technology, http://www.avere.org/.

ELEPHANTS

African Wildlife Foundation, http://www.awf.org.

U.S. Fish and Wildlife Service, Species List of Endangered and Threatened Wildlife, http://endangered.fws.gov/

World Wildlife Fund, http://www.wwf.org.

ETHANOL

U.S. Department of Energy, Energy Efficiency and Renewable Energy Clearinghouse,

P.O. Box 3048
Merrifield, VA 22116
E-mail: energyinfo@delphi.com.
Website: http://www.doe.gov.

EVERGLADES

National Park Service, Everglades National Park, http://www.nps.gov/ever.

FEDERAL EMERGENCY MANAGEMENT AGENCY (FEMA)

FEMA, http://www.fema.gov.

FISHING, COMMERCIAL

National Oceanographic and Atmospheric Administration Fisheries, http://www.nmfs. gov/.

United Nations Food and Agriculture Organization Fisheries, http://www.fao.org/waicent/faoinfo/fishery/fishery.htm.

FORESTS

American Forests, http://www.amfor.org.

Greenpeace International, Forests, http://www. greenpeace.org/~forests.

Society of American Foresters, http://www. safnet.org.

U.S. Forest Service, http://www.fs.fed.us.

U.S. Forest Service Research, http://www.fs.fed.us/links/research.shtml.

World Conservation Monitoring Centre, http://www.wcmc.org.uk.

World Resources Institute Forest Frontiers Initiative, http://www.wri.org/ffi.

World Wildlife Fund (Worldwide Fund for Nature) Forests for Life Campaign, http://www.panda.org/forests4life.

FUEL CELLS AND OTHER ALTERNATIVE FUELS

Crest's Guide to the Internet's Alternative Energy Resources, http://solstice.crest.org/online/aeguide/aehome.html.

U.S. Department of Energy

P.O. Box 12316
Arlington, VA 22209
Telephone: 1-800-423-1363
Website: http://www.doe.gov.

U.S. Department of Energy, Alternative Fuels Data Center, http://www.afdc.nrel.gov.

GEOLOGY

Geological surveys, U.S. Geological Survey, http://www.usgs.gov/.

For general interest publications and products, http://mapping.usgs.gov/www/products/mappubs.html.

GEOTHERMAL SITES

Energy and Geoscience Institute

University of Utah
423 Wakara Way
Salt Lake City, UT 84108
Website: http://www.egi.utah.edu.

Geothermal energy information, http://geothermal.marin.org.

Geothermal database USA and Worldwide, http://www.geothermal.org.

International geothermal, http://www.demon.co.uk/geosci/igahome.html.

Solstice is the Internet information service of the Center for Renewable Energy and Sustainable Technology (CREST), http://solstice.crest.org /

GLACIERS SHRINKING

United States Geological Survey, Climate Change and History, http://geology.usgs.gov/index. shtml.

Sierra Club, Public Information Center, (415) 923-5653; or the Global Warming and Energy Team, (202) 547-1141, or by E-mail: information@sierraclub.org.

GLOBEC

Educational Website, http://cbl.umces.edu/fogarty/usglobec/misc/education.html.

GRASSLANDS AND PRAIRIES

Postcards from the Prairie, http://www.nrwrc.usgs.gov/postcards/postcards.htm.

University of California, Berkeley, World Biomes, Grasslands, http://www.ucmp.berkeley.edu/glossary/gloss5/biome/grasslan.html.

Worldwide Fund for Nature, Grasslands and Its Animals, http://www.panda.org/kids/wildlife/idxgrsmn.htm.

GROUNDWATER

EPA, http://www.epa.gov/swerosps/ej/.

Groundwater atlas of the United States, http://www.capp.er.usgs.gov/publicdocs/gwa/.

HAZARDOUS MATERIALS TRANSPORTAION ACT

Website: http://www.dot.gov.

HAZARDOUS SUBSTANCES

U.S. Environmental Protection Agency Program, http://epa.gov/.

U.S. Occupational Safety and Health Administration (OSHA), http://www.osha.gov/toxicsubstances/index.html.

Environmental Defense Fund (data on wastes and chemicals at U.S. sources), http://www.scorecard.org.

HAZARDOUS WASTE TREATMENT

Federal Remedial Technologies Roundtable, Hazardous Waste Clean-Up Information ("CLU-IN"), http://www.clu-in.org.

HEAVY METALS

U.S. Environmental Protection Agency, Office of Pollution Prevention and Toxics, http://www.epa.gov/opptintr.

HIGH-LEVEL RADIOACTIVE WASTES

U.S. Nuclear Regulatory Commission, Radioactive Waste Page, http://www.nrc.gov/NRC/radwaste.

U.S. Environmental Protection Agency, Mixed-Waste Homepage, http://www.epa.gov/radiation/mixed-waste.

HURRICANES

National Hurricane Center, http://www.nhc.noaa.gov.

HYDROELECTRIC POWER

U.S. Bureau of Reclamation Hydropower Information, http://www.usbr.gov/power/edu/edu.htm.

U.S. Geological Survey, http://wwwga.usgs.gov/edu/hybiggest.html.

HYDROGEN

National Renewable Energy Laboratory, http://www.nrel.gov/lab/pao/hydrogen.html.

EnviroSource, Hydrogen InfoNet, http:///www.eren.doe.gov/hydrogen/infonet.html.

INTERNATIONAL ATOMIC ENERGY AGENCY

Agency, http://www.iaea.org.

Managing Radioactive Waste Fact Sheet, http://www.iaea.org/worldatom/inforesource/factsheets/manradwa.html.

INTERNATIONAL COUNCIL FOR LOCAL ENVIRONMENTAL INITIATIVES

Homepage, http://www.iclei.org.

INTERNATIONAL REGISTER OF POTENTIALLY TOXIC CHEMICALS

Homepage, http://www.unep.org/unep/program/hhwb/chemical/irptc/home.htm.

INTERNATIONAL WHALING COMMISSION

Homepage, http://www.ourworld.compuserve.com/homepages/iwcoffice.

INVERTEBRATES: THREATENED AND ENDANGERED

U.S. Fish and Wildlife Service, Species List of Endangered and Threatened Wildlife, http://endangered.fws.gov/

LANDSAT AND SATELLITE IMAGES

Earthshots, Satellite Images of Environmental Change, http://www.usgs.gov/Earthshots/.

Landsat Gateway, http://landsat.gsfc.nasa.gov/main.htm.

LEAD

National Lead Information Center's Clearinghouse, 1-800-424-LEAD, http://www.epa.gov/lead/.

LEOPARDS

U.S. Fish and Wildlife Service, Species List of Endangered and Threatened Wildlife, http://www.fws.gov/r9endspp/lsppinfo.html.

LITTER

Keep America Beautiful, http://www.kab.org.

MAMMALS

U.S. Fish and Wildlife Service, Vertebrate Animals, http://www.fws.gov/r9endspp/lsppinfo.html.

MANATEES

Save the Manatees, http://www.savethemanatee. org.

Sea World, Manatees, http://www.seaworld.org/manatee/sciclassman.html.

MARSHES

Environmental Protection Agency, Office of Wetlands, Oceans, Watersheds, http://www.epa.gov/owow/wetlands/wetland2.html.

North American Waterfowl and Wetlands Office, http://www.fws.gov/r9nawwo.

North American Wetlands Conservation Act, http://www.fws.gov/r9nawwo/nawcahp.html.

North American Wetlands Conservation Council, http://www.fws.gov/r9nawwo/nawcc.html.

Wetlands, wetlands-hotline@epamail.epa.gov.

MATERIAL SAFETY DATA SHEET

Toxic chemicals, http://www.siri.org/msds; http://www.ilpi.com/mads/index.html.

MENDES, CHICO

Chico Mendes, http://www.edf.org/chico.

NATURAL DISASTERS

Building Safer Structures, http://quake.wr.usgs. gov/QUAKES/FactSheets/SaferStructures/.

Center for Integration of Natural Disaster Information, http://cindi.usgs.gov/events/.

Earthquakes, http://quake.wr.usgs.gov/; http://geology.usgs.gov/quake.html. For the latest earthquake information http://quake.wr.usgs.gov/QUAKES/CURRENT/current.html

National Hurricane Center, http://www.nhc.noaa.gov.

U.S. Geological Survey, http://geology.usgs.gov/whatsnew.html.

NATIONAL MARINE FISHERIES

History of National Marine Fisheries Service, http://www.wh.whoi.edu/125th/history/century.html.

National Marine Fisheries, http://kingfish.ssp.nmfs.gov.

NOAA Fisheries, http://www.nmfs.gov/.

NATIONAL OCEAN AND ATMOSPHERIC ADMINISTRATION (NOAA)

Climate forecasting, http://www.cdc.noaa.gov/ Seasonal/.

El Niño Theme Page, http://www.pmel.noaa.gov/toga-tao/el-nino/nino-home-low.html.

Homepage, http://www.noaa.gov/.

Recover Protected Species, http://www.noaa.gov/nmfs/recover.html.

Safe Navigation Page, http://anchor.ncd.noaa.gov/psn/psn.htm.

NATIONAL WEATHER SERVICE

Homepage, http://www.nws.noaa.gov.

NATIONAL WILDLIFE REFUGE SYSTEM

Homepage, http://refuges.fws.gov/NWRSHomePage.html.

NATURAL GAS

American Gas Association, http://www.aga.org.

Oil and Gas Journal Online, http://www.ogjonline.com.

U.S. Department of Energy, Energy Information Administration, http://www.eia.doe.gov.

U.S. Department of Energy, Office of Fossil Energy, http://www.fe.doe.gov.

U.S. Geological Survey Energy, Resources Program, http://energy.usgs.gov/index.html.

NOISE POLLUTION

Noise Pollution Clearinghouse, http://www. nonoise.org.

NONPOINT SOURCES

Nonpoint Source Pollution Control Program, http://www.epa.gov/OWOW/NPS/whatudo.html; http://www.epa.gov/OWOW/ NPS/.

NUCLEAR ENERGY AND NUCLEAR REACTORS

American Nuclear Society, http://www.ans.org.

Nuclear Energy Institute, http://www.nei.org.

Nuclear Information and Resource Service, http://www.nirs.org.

U.S. Department of Energy, Office of Nuclear Energy, Science and Technology, http://www.ne.doe.gov.

U.S. Nuclear Regulatory Commission, http://www.nrc.gov.

NUCLEAR WASTE POLICY ACT

American Nuclear Society, http://www.ans.org.

Nuclear Energy Institute, http://www.nei.org.

NUCLEAR WASTE SITES

Hazard Ranking System, http://www.epa.gov/ superfund/programs/npl_hrs/hrsint.htm.

National Research Council, Board on Radioactive Waste Management, http://www4.nas.edu/brwm/brwm-res.nsf.

Superfund, http://www.pin.org/superguide.htm; http://www.epa.gov/superfund.

U.S. Department of Energy, Office of Civilian Radioactive Waste Management, http://www.rw.doe.gov.

U.S. Environmental Protection Agency, Mixed-Waste Homepage, http://www.epa.gov/radiation/mixed-waste.

U.S. Nuclear Regulatory Commission, Radioactive Waste Page, http://www.nrc.gov/ NRC/radwaste.

OCCUPATIONAL SAFETY AND HEALTH ACT (OSHA)

OSHA Homepage, http://www.osha.gov.

OCEAN THERMAL ENERGY CONVERSION (OTEC)

National Renewable Energy Laboratory

1617 Cole Boulevard
Golden, CO 80401
Website: http:llnrelinfo.nrel.gov.

Natural Energy Laboratory of Hawaii, http://bigisland.com/nelha/index.html.

OCEANS

National Oceanographic and Atmospheric Administration, http://www.noaa.gov/.

Safe Ocean Navigation Page, http://anchor.ncd.noaa.gov/psn/psn.htm.

OFFICE OF SURFACE MINING

Office of Surface Mining, http://www.osmre.gov.

Appalachian Clean Streams Initiative, majordomo@osmre.gov.

OLD-GROWTH FORESTS

Greenpeace International, Forests, http://www.greenpeace.org/~forests.

World Resources Institute, Forest Frontiers Initiative, http://www.wri.org/ffi.

OLMSTEAD, FREDERICK LAW

Homepage, http://fredericklawolmsted.com.

ORGANIZATION OF PETROLEUM EXPORTING COUNTRIES (OPRC)

Homepage, http://www.opec.org.

OVERFISHING

Information and data statistics, http://www.nmfs. gov.

National Aeronautics and Space Administration, Ocean Planet, http://seawifs.gsfc.nasa.gov/OCEAN_PLANET/HTML/peril_overfishing.html.

National Marine Fisheries Service, http://www. nmfs.gov.

NOAA, http://www.noaa.gov.

United Nations Food and Agricultural Organization, http://www.fao.org.

United Nations Food and Agriculture Organization Fisheries, http://www.fao.org/.

United Nations System, http://www.unsystem.org.

OZONE-RELATED ISSUES

Environmental Protection Agency, science of ozone depletion, http://www.epa.gov/ozone/science/.

NOAA, Commonly Asked Questions about Ozone, www.publicaffairs.noaa.gov/grounders/ozo1.html.

NOAA, Network for the Detection of Stratospheric Change, www.noaa.gov.

PARROTS

Online Book of Parrots, http://www.ub.tu-clausthal.dep/p_welcome.html.

World Parrot Trust, http://www.worldparrottrust.org.

World Wildlife Fund, http:www.panda.org.

PESTICIDES

Toxics and Pesticides, http://www.epa.gov/oppfead1/work_saf/.

Pesticides in the Atmosphere, http://ca.water.usgs.gov/pnsp/atmos.

PETERSON, ROGER TORY

Roger Tory Peterson Institute of Natural History,

311 Curtis Street
Jamestown, NY 14701
Website: http://www.rtpi.org/info/rtp.htm.

PETROLEUM

American Petroleum Institute, http://www.api.org.

Petroleum Information, http://www.petroleuminformation.com.

Oil and Gas Journal Online, http://www. ogjonline.com.

U.S. Department of Energy, Energy Information Administration, http://www.eia.doe.gov.

U.S. Department of Energy, Office of Fossil Energy, http://www.fe.doe.gov.

U.S. Geological Survey Energy Resources Program, http://energy.usgs.gov/index.html.

U.S. Geological Survey Fact Sheet FS-145-97, "Changing Perceptions of World Oil and Gas Resources as Shown by Recent USGS Petroleum Assessments," http://greenwood.cr.usgs.gov/pub/fact-sheets/fs-0145-97/fs-0145-97.html.

PLUTONIUM

U.S. Nuclear Regulatory Commission, Radioactive Waste Page, http://www.nrc.gov/NRC/radwaste.

RADIATION AND RADIOACTIVE WASTES

International Atomic Energy Agency, "Managing Radioactive Waste" Fact Sheet, http://www.iaea.org/worldatom/inforesource/factsheets/manradwa.html.

National Research Council, Board on Radioactive Waste Management, http://www4.nas.edu/brwm/brwm-res.nsf.

U.S. Department of Energy, Office of Civilian Radioactive Waste Management, http://www. rw.doe.gov.

U.S. Environmental Protection Agency, Mixed-Waste Homepage, http://www.epa.gov/radiation/mixed-waste.

U.S. Nuclear Regulatory Commission, Radioactive Waste Page, http://www.nrc.gov/NRC/radwaste.

RADON

Radon in Earth, Air, and Water, http://sedwww.cr.usgs.gov:8080/radon/radonhome.html.

RAIN FORESTS

Greenpeace International, forests, http://www.greenpeace.org/~forests.

Rainforest Action Network (RAN)

President Randy Hayes
221 Pine Street Suite 500
San Francisco, CA 94104
Telephone: (415) 398-4404
Website: http://www.ran.org

Rainforest Alliance (RA)

65 Bleeker Street
New York, NY 10012
Website: http://www.rainforest-alliance.org

U.S. Forest Service, http://www.fs.fed.us.

World Wildlife Fund (Worldwide Fund for Nature), Forests for Life Campaign, http://www.panda.org/forests4life.

RESOURCE CONSERVATION AND RECOVERY ACT

Homepage, http://www.epa.gov/epaoswer/hotline.

SALMON

National Marine Fisheries Service, http://www.nwr.noaa.gov/1salmon/salmesa/index.htm.
NOAA Fisheries, http://www.nmfs.gov/.

SALT MARSHES

National Wetlands Research Center, http://www.nwrc.usgs.gov/educ_out.html.

USGS Coastal and Marine Geology, http://marine.usgs.gov/.

SANITARY LANDFILLS

Solid waste management, http://web.mit.edu/urbanupgrading/urban environment/

Landfills - Solid and Hazardous Waste and Groundwater Quality Protection, http://www.gfredlee.com/plandfil2.htm

SIBERIA

Siberia, http://www.cnit.nsk.su/univer/english/siberia.htm.

SOLAR ENERGY

American Solar Energy Society

2400 Central Avenue, Suite G-1
Boulder, CO 80301.
Website: http://www.soton.ac.uk/~solar/.

Solar Energy Industries Association

122 C Street, NW, 4th Floor
Washington, D.C. 20001.
Website: http://www.seia.org/main.htm.

U.S. Department of Energy, Photovoltaic Program, http://www.eren.doe.gov/pv/text_frameset.html.

SOLAR POND

Department of Mechanical and Industrial Engineering

University of Texas at El Paso
El Paso, TX 79968.
E-mail: aswift@cs.utep.edu.

SPENT FUEL

Environmental Protection Agency, www.ntp.doe.gov, www.rw.doe.gov/pages/resource/facts/transfct.htm.

SUPERFUND

Environmental Protection Agency, http://www.epa.gov/epaoswer/hotline.

Recycled Superfund sites, http://www.epa.gov/superfund/programs/recycle/index.htm.

Superfund Information, http://www.epa.gov/superfund.

U.S. EPA Superfund Program Homepage, Website: http://www.epa.gov/superfund/index.htm.

TENNESSEE VALLEY AUTHORITY

Homepage, http://www.tva.gov.

THOREAU, HENRY

Website: http://www.walden.org.

TOXIC CHEMICALS

Environmental Defense Fund, http://www.scorecard.org.

U.S. Department of Health and Human Services, Agency for Toxic Substances and Disease Registry (ASTDR), http://www.atsdr.cdc.gov/

U.S. Environmental Protection Agency, Integrated Risk Information System (IRIS), http://www. siri.org/msds; http://www.ilpi.com/mads/index.html.

U.S. Occupation Health and Safety Administration, http://www.toxicsubstances/index.html.

TOXIC RELEASE INVENTORY

Environmental Defense Fund, http://www.scorecard.org.
Environmental Protection Agency, http://www.epa.gov.

Teach with Databases, Toxic Release Inventory, http://www.nsta.org/pubs/special/pb143x01.htm.

TOXIC WASTE

Environmental Defense Fund, http://www.scorecard.org.

Institute for Global Communications, http://www.igc.org/igc/issues/tw/.

TRADE RECORDS ANALYSIS OF FLORA AND FAUNA IN COMMERCE (TRAFFIC)

Homepage, http://www.traffic.org/about/.

URBAN FORESTS

American Forests, http://www.amfor.org.

TreeLink, http://www.treelink.org.

VERTEBRATES

U.S. Fish and Wildlife Service, Species List of Endangered and Threatened Wildlife, http://www.fws.gov/r9endspp/lsppinfo.html.

VICUNA

U.S. Fish and Wildlife Service, Species List of Endangered and Threatened Wildlife, http://endangered.fws.gov

VITRIFICATION

U.S. Department of Energy, http://www.em.doe.gov/fs/fs3m.html.

VOLCANOES

USGS, Volcanoes in the Learning Web, http://www.usgs.gov/education/learnweb/volcano/index.html.

Volcano Hazards, http://volcanoes.usgs.gov/.

WATER CONSERVATION AND POLLUTION

Early History of the Clean Water Act, http://epa.gov/history/topics.

Environmental Protection Agency, Office of Wetlands, Oceans, Watersheds for Nonpoint Source information, http://www.epa.gov/owow/wetlands/wetland2.html; http://www.epa.gov/swerosps/ej/.

U.S. Geological Survey, Water Resources of the United States, National Groundwater Association Homepage, http://www.h2o-ngwa.org.

Water Resources Information, http://water.usgs.gov/.

Water Use Data, http://water.usgs.gov/public/watuse/.

WETLANDS

National Wetlands Research Center, http://www.nwrc.usgs.gov/educ_out.html.

Ramsar Convention on Wetlands (International), http://www2.iucn.org/themes/ramsar/.

Ramsar List of Wetlands of International Importance, http://ramsar.org/key_sitelist.htm.

WHALES

Institute of Cetacean Research (ICR), http://www.whalesci.org.

U.S. Fish and Wildlife Service, Species List of Endangered and Threatened Wildlife, http://www.fws.gov/r9endspp/lsppinfo.html; http://www.highnorth.no/iceland/th-in-to.htm; http://greenpeace.org/.

WILDERNESS

U.S. Forest Service, *Roadless Area Review and Evaluation*, http://www.fs.fed.us.

Wilderness Society, http://www.wilderness.org/newsroom/factsheets.htm.

WILDLIFE REFUGES

Conservation International, http://www.conservation.org.

Nature Conservancy, http://www.tnc.org.

U.S. Fish and Wildlife Service, National Wildlife Refuge System, http://refuges.fws.gov.

World Conservation Union/International Union for the Conservation of Nature, http://www.iucn.org.

WIND ENERGY

American Wind Energy Association

122 C Street NW, 4th Floor
Washington, D.C. 20001
Telephone: (202) 383-2500.
E-mail: awea@mcimail.com.
Website: http://www.awea.org.

Center for Renewable Energy and Sustainable Technology (CREST)

Solar Energy Research and Education Foundation
777 North Capitol Street NE, Suite 805
Washington, D.C. 20002
Website: http://solstice.crest.org/.

WOLVES

U.S. Fish and Wildlife Service, http://www.fws.gov/.
U.S. Fish and Wildlife Service, Species List of Endangered and Threatened Wildlife, http://endangered.fws.gov/.

World Wildlife Fund

1250 24th Street, NW
Washington, D.C. 20037
Telephone: 1-800-225-5993
Website: http://www.worldwildlife.org/.

WORLD HEALTH ORGANIZATION

Homepage, http://www.who.int.

WORLD WILDLIFE FUND

1250 24th Street, NW
Washington, D.C. 20037
Telephone: 1-800-225-5993
Website: http://www.wwf.org/.

YUCCA MOUNTAIN PROJECT

Homepage, http://www.ymp.gov/.

ZEBRAS

U.S. Fish and Wildlife Service, Species List of Endangered and Threatened Wildlife, http://endangered.fws.gov/.

ZOOS

Bronx Zoo, http://www.bronxzoo.com/.
San Diego Zoo, http://www.sandiegozoo.org/.

Appendix D: Environmental Organizations

Action for Animals

P.O. Box 17702
Austin, TX 78760
Telephone: (512) 416-1617
Fax: (512) 445-3454
Website: http://www.envirolink.org/

African Wildlife Foundation (AWF)

1400 Sixteenth Street, NW, Suite 120
Washington, D.C. 20036
Telephone: (202) 939-3333
Fax: (202) 939-3332
Website: http://www.awf.org/home.html

Agency for Toxic Substances and Diseases, Registry Division of Toxicology (ATSDR)

1600 Clifton Road
NE Mailstop E-29
Atlanta, GA 30333
Telephone: (888) 42-ATSDR or (888) 422-8737
E-mail: ATSDRIC@cdc.gov
Website: http://www.atsdr.cdc.gov/contacts.html

Alaska Forum for Environmental Responsibility

P.O. Box 188
Valdez, AK 99686
Telephone: (907) 835-5460
Fax: (907) 835-5410
Website: http://www.accessone.com/~afersea

American Conifer Society (ACS)

P.O. Box 360
Keswick, VA 22947-0360
Telephone: (804) 984-3660
Fax: (804) 984-3660
E-mail: ACSconifer@aol.com
Website: http://www.pacificrim.net/~bydesign/acs.html

American Forests

P.O. Box 2000
Washington, D.C. 20013
Telephone: (202) 955-4500
Website: http://www.americanforests.org

American Nuclear Society

555 North Kensington Avenue
La Grange Park, IL 60525
Telephone: (708) 352-6611
Fax: (708) 352-0499
E-mail: NUCLEUS@ans.org
Website: http://www.ans.org

American Oceans Campaign

201 Massachusetts Avenue NE, Suite C-3
Washington, D.C. 20002
Telephone: (202) 544-3526
Fax: (202) 544-5625
E-mail: aocdc@wizard.net
Website: http://www.americanoceans.org

American Rivers

1025 Vermont Avenue NW, Suite 720
Washington, D.C. 20005
Telephone: (202) 347-7500
Fax: (202) 347-9240
E-mail: amrivers@amrivers.org
Website: http://www.amrivers.org

American Society for Horticultural Science (ASHS)

600 Cameron Street
Alexandria, VA 22314-2562

Telephone: (703) 836-4606
Fax: (703) 836-2024
E-mail: webmaster@ashs.org
Website: http://www.ashs.org

American Society for the Prevention of Cruelty to Animals (ASPCA)

424 East Ninety-second Street
New York, NY 10128
Telephone: (212) 876-7700
Website: http://www.aspca.org

American Solar Energy Society

2400 Central Avenue, Suite G-1
Boulder, CO 80301
Telephone: (303) 443-3130
Fax: (303) 443-3212
E-mail: ases@ases.org
Website: http://www.ases.org
Publication: *Solar Today*

American Wind Energy Association

122 C Street NW, Fourth Floor
Washington, D.C. 20001
Telephone: (202) 383-2500
E-mail: awea@mcimail.com
Website: http://www.awea.org

Animal Legal Defense Fund (ALDF)

127 Fourth Street
Petaluma, CA 94952
Telephone: (707) 769-7771
Fax: (707) 769-0785
E-mail: info@aldf.org
Website: http://www.aldf.org

Animal Rights Network

P.O. Box 25881
Baltimore, MD 21224
Telephone: (410) 675-4566
Fax: (410) 675-0066
Website: http://www.envirolink.org/arrs/aa/index.html
Publication: *Animals' AGENDA*, a bimonthly magazine

Baron's Haven Freehold

104 South Main Street
Cadiz, OH 43907
Telephone: (740) 942-8405
Website: http://bhfi.1st.net

Biodiversity Support Program (BSP)

1250 North Twenty-fourth Street NW, Suite 600
Washington, D.C. 20037
Telephone: (202) 778-9681
Fax: (202) 861-8324
Website: http://www.BSPonline.org

Biosfera

Pres. Vargas 435, Suites 1104 and 1105
Rio de Janeiro, RJ 20077-900
Brazil

Birds of Prey Foundation

2290 South 104th Street
Broomfield, CO 80020
Telephone: (303) 460-0674
Fax: (303) 666-1050
E-mail: raptor@birds-of-prey.org
Website: http://www.birds-of-prey.org

Build the Earth

3818 Surfwood Road
Malibu, CA 90265
Telephone: (310) 454-0963

Center for Conversion and Research of Endangered Wildlife (CREW)

Cincinnati Zoo and Botanical Garden
3400 Vine Street
Cincinnati, OH 45220
E-mail: terri.roth@cincyzoo.org

Center for Marine Conservation

1725 DeSales Street SW, Suite 600
Washington, D.C. 20036
Telephone: (202) 429-5609
Fax: (202) 872-0619
E-mail: cmc@dccmc.org
Website: http://www.cmc-ocean.org

Centers for Disease Control (CDC)

1600 Clifton Rd.
Atlanta, GA 30333

Telephone: (800) 311-3435
Website: http://www.cdc.gov

Cheetah Conservation Fund (CCF)

P.O. Box 1380
Ojai, CA 93024
Telephone: (805) 640-0390
Fax: (815) 640-0230
E-mail: info@cheetah.org
Website: http://www.cheetah.org

Clean Air Council (CAC)

135 South Nineteenth Street, Suite 300
Philadelphia, PA 19103
Telephone: (888) 567-7796
Website: http://www.libertynet.org/~cleanair/

Coalition for Economically Responsible Economies (CERES)

11 Arlington Street, Sixth Floor
Boston, MA 02116-3411
Telephone: (617) 247-0700
Fax: (617) 267-5400
Website: http://www.ceres.org

Conservation International

1015 Eighteenth Street NW Suite 1000
Washington, D.C. 20036
Telephone: (202) 429-5660
Website: http://www.conservation.org/
Publication: *Orion Nature Quarterly*

Convention on International Trade in Endangered Species of Wild Fauna and Flora (CITES)

CITES Secretariat
International Environment House, 15, chemin des Anémones, CH-1219
Châtelaine-Geneva, Switzerland
E-mail: cites@unep.ch
Website: http://www.cites.org/index.shtml

Council for Responsible Genetics

5 Upland Road, Suite 3
Cambridge, MA 02140
Website: http://www.gene-watch.org

Cousteau Society

870 Greenbriar Circle, Suite 402
Chesapeake, VA 23320
Telephone: (804) 523-9335
E-mail: cousteau@infi.net
Website: http://www.cousteausociety.org/
Publication: *Calypso Log*

Defenders of Wildlife

1101 Fourteenth Street NW, Room 1400
Washington, D.C. 20005
Telephone: (800) 441-4395
Website: http://www.Defenders.org
Publication: *Defenders*, a quarterly magazine

Dian Fossey Gorilla Fund International

800 Cherokee Avenue SE
Atlanta, GA 30315-1440
Telephone: (800) 851-0203
Fax: (404) 624-5999
E-mail: 2help@gorillafund.org
Website: http://www.gorillafund.org/000_core_frmset.html

Earth Day Network

1616 P Street NW
Suite 200
Washington, D.C. 20036
E-mail: earthday@earthday.net
Website: http://www.earthday.net

Earth Island Institute (EII)

300 Broadway, Suite 28
San Francisco, CA 94133
Telephone: (415) 788-3666
Fax: (415) 788-7324
Website: http://www.earthisland.org/abouteii/abouteii.html
Publication: *Earth Island Journal*, a quarterly magazine

Earth, Pulp, and Paper

P.O. Box 64
Leggett, CA 95585
Telephone: (707) 925-6494
E-mail: tree@tree.org
Website: http://www.tree.org/epp.htm

EarthFirst! (EF!)

P.O. Box 5176
Missoula, MT 59806
Website: http://www.webdirectory.com/General_Environmental_Interest/Earth_First_/

Earthwatch Institute

In United States and Canada
3 Clocktower Place, Suite 100
Box 75
Maynard, MA 01754
Telephone: (800) 776-0188 or (617) 926-8200
Fax: (617) 926-8532
In Europe
57 Woodstock Road
Oxford OX2 6HJ, United Kingdom
E-mail: info@uk.earthwatch.org
Website: http://www.earthwatch.org

EcoCorps

1585 A Folsom Avenue
San Francisco, CA 94103
Telephone: (415) 522-1680
Fax: (415) 626-1510
E-mail: eathvoice@ecocorps.org
Website: http://www.owplaza.com/eco

Ecotourism Society

P.O. Box 755
North Bennington, VT 05257
Telephone: (802) 447-2121
Fax: (802) 447-2122
E-mail: ecomail@ecotourism.org
Website: http://www.ecotoursim.org

E. F. Schumacher Society

140 Jug End Road
Great Barrington, MA 01230
Telephone: (413) 528-1737
E-mail: efssociety@aol.com
Website: http://members.aol.com/efssociety/index.html

Electric Vehicle Association of the Americas

701 Pennsylvania Avenue NW, Fourth Floor
Washington, D.C. 20004
Telephone: (202) 508-5995
Fax: (202) 508-5924
Website: http://www.evaa.org

Environmental Defense Fund (EDF)

257 Park Avenue South
New York, NY 10010
Telephone: (800) 684-3322
Fax: (212) 505-2375
E-mail (for general questions and information): Contact@environmentaldefense.org
Website: http://www.edf.org
Publication: *Nature Journal*, a monthly magazine

Exotic Cat Refuge and Wildlife Orphanage

Route 3, Box 96A
Kirbyville, TX 75956
Telephone: (409) 423-4847

Federal Emergency and Management Agency (FEMA)

500 C Street SW
Washington, D.C. 20472
Website: http://www.fema.gov

Friends of the Earth (FOE)

1025 Vermont Avenue NW, Suite 300
Washington, D.C. 20005-6303
Telephone: (202) 783-7400
Fax: (202) 783-0444
E-mail: foe@foe.org
Website: http://www.foe.org

Green Seal

1001 Connecticut Avenue NW, Suite 827
Washington, D.C. 20036-5525
Telephone: (202) 872-6400
Fax: (202) 872-4324
Website: http://www.greenseal.org

Greenpeace USA

1436 U Street NW
Washington, D.C. 20009
Telephone: (202) 462-1177
Website: http://www.greenpeaceusa.org/
Publication: *Greenpeace Magazine*

Hawkwatch International

P.O. Box 660
Salt Lake City, UT 84110
Telephone: (801) 524-8511
E-mail: hawkwatch@charitiesusa.com
Website: http://www.vpp.com/HawkWatch

Humane Society of the United States (HSUS)

2100 L Street NW
Washington, D.C. 20037
Website: http://www.hsus.org
Publications: *All Animals*, a quarterly magazine

International Atomic Energy Commission

P.O. Box 100
Wagramer Strasse 5
A-1400, Vienna, Austria
E-mail: Official.Mail@iaea.org
Website: http://www.iaea.org

International Council for Local Environmental Initiatives (ICLEI)

World Secretariat
16th Floor, West Tower, City Hall
Toronto, M5H 2N2, Canada
Fax: (416) 392-1478
Email: iclei@iclei.org
Website: http://www.iclei.org

International Rhino Foundation (IRF)

14000 International Road
Cumberland Ohio 43732
E-mail: IrhinoF@aol.com
Website: http://www.rhinos-irf.org

International Whaling Commission (IWC)

The Red House
135 Station Road
Impington, Cambridge CB4 9NP, United Kingdom
E-mail: iwc@iwcoffice.org
Website: http://ourworld.compuserve.com/homepages/iwcoffice

International Wolf Center

1396 Highway 169
Ely, MN 55731-8129
Telephone: (218) 365-4695
Fax: (218) 365-3318
Website: http://www.wolf.org

Jane Goodall Institute (JGI)

P.O. Box 14890
Silver Spring, MD 20911-4890
Telephone: (301) 565-0086
Fax: (301) 565-3188
E-mail: JGIinformation@janegoodall.org

Keep America Beautiful

1010 Washington Boulevard
Stamford, CT 06901
Telephone: (203) 323-8987
Fax: (203) 325-9199
E-mail: info@kab.org

League of Conservation Voters

1707 L Street, NW, Suite 750
Washington, D.C. 20036
Telephone: (202) 785-8683
Fax: (202) 835-0491
E-mail: lcv@lcv.org
Website: http://www.lcv.org

Mountain Lion Foundation (MLF)

P.O. Box 1896
Sacramento, CA 95812
Telephone: (916) 442-2666
E-mail: MLF@moutainlion.org
Website: http://www.mountainlion.org

National Alliance of River, Sound, and Bay Keepers

P.O. Box 130
Garrison, NY 10524
Telephone: (800) 217-4837
E-mail: keepers@keeper.org
Website: http://www.keeper.org

National Anti-Vivisection Society (NAVS)

53 West Jackson Street, Suite 1552
Chicago, IL 60604
Telephone: (800) 888-NAVS
E-mail: navs@navs.org
Website: http://www.navs.org

National Arbor Day Foundation

100 Arbor Avenue
Nebraska City, NE 68410
Telephone: (402) 474-5655
Website: http://www.arborday.org
Publication: *Arbor Day*, a bimonthly magazine

National Audubon Society (NAS)

700 Broadway
New York, NY 10003
Telephone: (212) 979-3000
Website: http://www.audubon.org
Publication: *Audubon*, a bimonthly magazine

National Center for Environmental Health

Mail Stop F-29
4770 Buford Highway NE
Atlanta, GA 30341-3724
Telephone NCEH Health Line: (888) 232-6789
Website: http://www.cdc.gov/nceh/ncehhome.htm

National Parks and Conservation Association (NPCA)

1015 Thirty-first Street NW
Washington, D.C. 20007
Telephone: (202) 944-8530; (800) NAT-PARK
E-mail: npca@npca.org
Website: http://www.npca.org
Publication: *National Parks*, a bimonthly magazine

National Wildlife Federation (NWF)

8925 Leesburg Pike
Vienna, VA 22184-0001
Telephone: (800) 822-9919
Website: http://www.nwf.org
Publication: *National Wildlife*, a bimonthly magazine

Natural Resources Defense Council (NRDC)

40 West Twentieth Street
New York, NY 10011
Website: http://www.nrdc.org
Publications: *Amiscus Journal*, a quarterly magazine

Nature Conservancy (TNC)

1815 North Lynn Street
Arlington, VA 22209
Telephone: (703) 841-5300
Fax: (703) 841-1283
Website: http://www.tnc.org
Publication: *Nature Conservancy*, a magazine

Noise Pollution Clearinghouse

P.O. Box 1137
Montpelier, VT 05601-1137
Telephone: (888) 200-8332
Website: http://www.nonoise.org

North Sea Commission

Business and Development Office
Skottenborg 26, DK-8800 Viborg, Denmark
Website: http:\\www.northsea.org

People for Animal Rights

P.O. Box 8707
Kansas City, MO 64114
Telephone: (816) 767-1199
E-mail: parinfo@envirolink.org
Website: http://www.parkc.org

People for the Ethical Treatment of Animals (PETA)

501 Front Street
Norfolk, VA 23510
Telephone: (757) 622-PETA
Fax: (757) 622-0457
Website: http://www.peta-online.org/

Orangutan Foundation International

822 South Wellesley Avenue
Los Angeles, CA 90049
Telephone: (800) ORANGUTAN
Fax: (310) 207-1556
E-mail: ofi@orangutan.org
Website: http://www.ns.net/orangutan

Ozone Action

1700 Connecticut Avenue NW, Third Floor
Washington, D.C. 20009
Telephone: (202) 265-6738

E-mail: cantando@essential.org
Website: www.ozone.org

Peregrine Fund

566 West Flying Hawk Lane
Boise, ID 83709
Telephone: (208) 362-3716
Fax: (208) 362-2376
E-mail: tpf@peregrinefund.org
Website: http://www.peregrinefund.org

Rachel Carson Council

8940 Jones Mill Road
Chevy Chase, MD 20815
Telephone: (301) 652-1877
E-mail: rccouncil@aol.com
Website: http://members.aol.com/rccouncil/ourpage

Rainforest Action Network

221 Pine Street, Suite 500
San Francisco, CA 94104-2740
Telephone: (415) 398-4404
Fax: (415) 398-2732
E-mail: rainforest@ran.org
Website: http://www.ran.org

Range Watch

45661 Poso Park Drive
Posey, CA 93260
Telephone: (805) 536-8668
E-mail: rangewatch@aol.com
Website: http://www.rangewatch.org

Raptor Resource Project

2580 310th Street
Ridgeway, IA 52165
E-mail: rrp@salamander.com
Website: http://www.salamander.com~rpp

Reef Relief

201 William Street
Key West, FL 33041
Telephone: (305) 294-3100
Fax: (305) 923-9515
E-mail: reef@bellsouth.net
Website: http://www.reefrelief.org

ReefKeeper International

2809 Bird Avenue, Suite 162
Miami, FL 33133
Telephone: (305) 358-4600
Fax: (305) 358-3030
E-mail: reefkeeper@reefkeeper.org
Website: http://www.reefkeeper.org

Renewable Energy Policy Project-Center for Renewable Energy and Sustainable Technology (REPP-CREST)

National Headquarters
1612 K Street, NW, Suite 202
Washington, D.C. 20006
Website: http://www.solstice.crest.org

Resources for the Future (RFF)

1616 P Street NW
Washington, D.C. 20036
Telephone: (202) 328-5000
Fax: (202) 939-3460
E-mail: info@rff.org
Website: http://www.rff.org

Roger Tory Peterson Institute

311 Curtis Street
Jamestown, NY 14701
Telephone: (716) 665-2473
E-mail: webmaster@rtpi.org

Sierra Club

85 Second Street, Second Floor
San Francisco, CA 94105
Telephone: (415) 977-5630
Fax: (415) 977-5799
E-mail (general information): information@sierraclub.org
Website: http://www.Sierraclub.org
Publication: *Sierra*, a bimonthly magazine

Smithsonian Institution Conservation & Research Center (CRC)

Website: http://www.si.edu/crc/brochure/index.htm

Society of American Foresters

5400 Grosvenor Lane
Bethesda, MD 20814

Telephone: (301) 897-8720
Fax: (301) 897-3690
E-mail: safweb@safnet.org
Website: http://www.safnet.org

Surfrider Foundation USA

122 South El Camino Real, Suite 67
San Clemente, CA 92672
Telephone: (949) 492-8170
Fax: (949) 492-8142
Website: http://www.surfrider.org

Union of Concerned Scientists

National Headquarters
2 Brattle Square
Cambridge, MA 02238
Telephone: (617) 547-5552
E-mail: ucs@ucsusa.org
Website: http://www.ucsusa.org
Publications: *Nucleus*, a quarterly magazine; *Earthwise*, a quarterly newsletter

United Nations Environment Programme (Regional)

2 United Nations Plaza
NY, NY 10017
Telephone: (212) 963-8138
Website: http://www.unep.org

United Nations Food and Agriculture Organization (FAO)

Website: http://www.fao.org
Liaison office with North America
Suite 300, 2175 K Street NW, Washington D.C. 20437-0001

United Nations Man and the Biosphere Programme (UNMAB)

U.S. MAB Secretariat, OES/ETC/MAB
Department of State
Washington, D.C. 20522-4401
Website: http://www.mabnet.org

U.S. Department of Agriculture (USDA)

14th Street and Independence Avenue., SW,
Washington, D.C. 20250
Website: http://www.usda.gov

U.S. Department of Energy (DOE)

Forrestal Building
1000 Independence Avenue, SW,
Washington, D.C. 20585
Website: http://www.doe.gov

U.S. Environmental Protection Agency (EPA)

401 M Street SW
Washington, D.C. 20460
Website: http://www.epa.gov

U.S. Fish and Wildlife Service (FWS)

1849 C Street NW
Washington, D.C. 20240
Telephone: (202) 208-5634
Website: http://www.fws.org

U.S. Geological Survey (USGS)

U.S. Dept. of Interior
1849 C Street, NW
Washington, D.C. 20240
Website: http://www.usgs.gov

U.S. National Park Service (NPS)

U.S. Dept. of Interior
1849 C Street, NW
Washington, D.C. 20240
Website: http://www.nps.gov

U.S. Nuclear Regulatory Commission (NRC)

One White Flint North
11555 Rockville Pike
Rockville, Maryland 20852
Website: http://www.nrc.gov

Wilderness Society

900 Seventeenth Street NW
Washington, D.C. 20006-2506
Telephone: (800) THE-WILD
Website: www.wilderness.org

Wildlands Project (TWP)

1955 West Grant Road, Suite 145
Tucson, AZ 85745
Telephone: (520) 884-0875
Fax: (520) 884-0962

E-mail: information@twp.org
Website: http://www.twp.org

World Conservation Monitoring Centre (WCMC)

219 Huntington Road
Cambridge CB3 ODL, United Kingdom
E-mail: info@wcmc.org.uk
Website: http://www.wcmc.org.uk

World Conservation Union (IUCN)

1630 Connecticut Avenue NW, Third Floor
Washington, D.C. 20009-1053
Telephone: (202) 387-4826
Fax: (202) 387-4823
E-mail: postmaster@iucnus.org
Website: http://www.iucn.org

World Health Organization (WHO)

Avenue Appia 20
1211 Geneva 27
Switzerland
Website: http://www.eho.int
E-mail: inf@who.int

World Parrot Trust United States

P.O. Box 50733
Saint Paul, MN 55150
Telephone: (651) 994-2581
Fax: (651) 994-2580
E-mail: usa@worldparrottrust.org

United Kingdom

Karen Allmann, Administrator,
Glanmor HouseHayle,
Cornwall TR27 4HY,
United Kingdom
E-mail: uk@worldparrottrust.org

Australia

Mike Owen
7 Monteray Street
Mooloolaba, Queensland 4557, Australia
E-mail: australia@worldparrottrust.org
Website: http://www.world parrottrust.org

World Resources Institute

1709 New York Avenue NW
Washington, D.C. 20006
Telephone: (202) 638-6300
E-mail: info@wri.org
Website: http://www.wri.org/wri/biodiv

World Society for the Protection of Animals (WSPA)

P.O. Box 190
Jamaica Plain, MA 02130
Website: http://www.wspa.org

United Kingdom Division
Website: http://www.wspa.org.uk/home.html

World Wildlife Fund, US (WWF)

1250 Twenty-fourth Street NW
P.O. Box 97180
Washington, D.C. 20077-7180
Telephone: (800) CALL-WWF
Website: http://www.worldwildlife.org

WorldWatch Institute

1776 Massachusetts Avenue NW
Washington, D.C. 20036
Telephone: (202) 452-1999
Website: http://www.worldwatch.org/
Publications: *WorldWatch, State of the World, Vital Signs* (annuals)

Zero Population Growth

1400 Sixteenth Street NW, Suite 320
Washington, D.C. 20036
Telephone: (202) 332-2200
Fax: (202) 332-2302
E-mail: zpg@igc.apc.org
Website: http://www.zpg.org

Zoe Foundation

983 River Road
Johns Island, SC 29455
Telephone: (803) 559-4790
E-mail: savage@awod.com
Website: http://www.2zoe.com

Index

f indicates figures and photos; t indicates tables

1: Earth Systems, Volume I **2:** Resources, Volume II **3:** People, Volume III **4:** Human Impact, Volume IV **5:** Sustainable Society, Volume V

1: Earth Systems, Volume I
2: Resources, Volume II
3: People, Volume III
4: Human Impact, Volume IV
5: Sustainable Society, Volume V

1: Earth Systems, Volume I
2: Resources, Volume II
3: People, Volume III
4: Human Impact, Volume IV
5: Sustainable Society, Volume V

1: Earth Systems, Volume I
2: Resources, Volume II
3: People, Volume III
4: Human Impact, Volume IV
5: Sustainable Society, Volume V

1: Earth Systems, Volume I
2: Resources, Volume II
3: People, Volume III
4: Human Impact, Volume IV
5: Sustainable Society, Volume V

ABOUT THE AUTHORS

JOHN MONGILLO is a noted science writer and educator. He is coauthor of *Encyclopedia of Environmental Science*, and *Environmental Activists*, both available from Greenwood.

PETER MONGILLO has won several awards for his teaching, including School District Teacher of the Year, National Endowment for the Humanities Fellowship Award, and the National Council for Geographic Education Distinguished Teacher Award.